韩国建筑世界出版社 编

亚洲 ASIA

155个居住设计 HOUSING 155

公寓 APARTMENT

下 册

大连理工大学出版社

目录 Contents

出版人的话
Publisher's message

The lexical meaning of housing is a building that protects men from natural harms such as rain, wind, cold and heat and social harms such as theft and destruction. It is not an over-statement to say that human history had begun together with housing. Men cannot live without housing.

Being faithful to basic is always the most important. Therefore, I believe it is quite meaningful to collect the housing of the world, which is the foundation of architecture and the essence of living, and introduce them to the readers.

The five Olympic rings which symbolize five continents are the motif of this book. In the Olympic rings, the blue, yellow, black, green and red rings from left are overlapped in W-shape. Likewise, this book introduces the multi-family housing recently completed in Europe, America, Oceania and Asia. The concept of family is also changing now and this book tries to suggest the future direction of housing by showing the rapidly changing concepts on housing and residence by way of showing recently completed architectural works.

I sincerely hope that "I·HOUSING" will become one of the bases in learning the housing trend in Korea and overseas to the readers. It would be my pleasure if readers would get useful information and wise insight from this book.

As last, I would like to express my sincere appreciation to the architects and owners who helped us with information and photographing opportunities.

I promise that the Archiworld will continue to provide dear readers with solid contents and great editing. Meanwhile, your continuous interest and support would be highly appreciated.

JEONG, KWANG YOUNG
Publisher

亚洲 ASIA

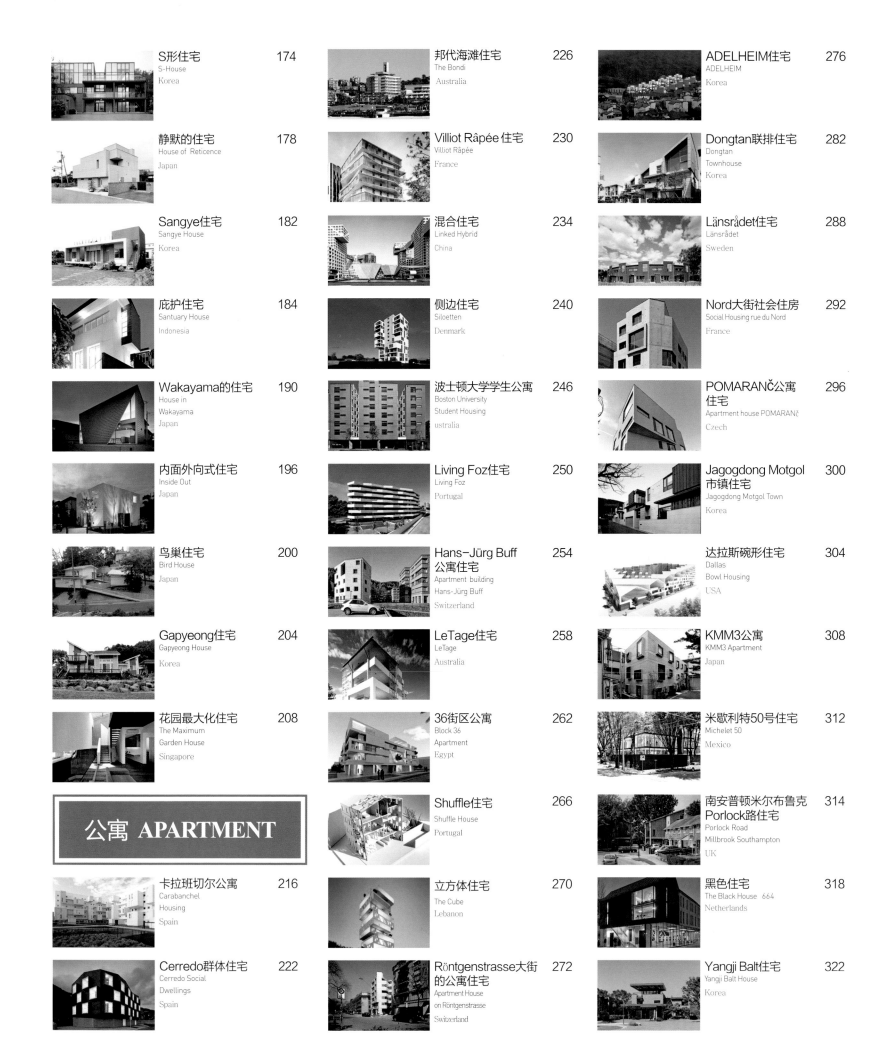

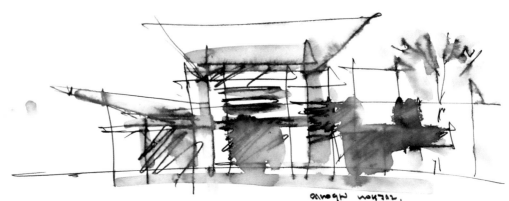

亚洲 ASIA

折纸住宅 Origami House

Location Singapore
Site area 558㎡
Gross floor area 622㎡
Architecture design Formwerkz
Architects
Design participation Gwen Tan Tze Suen,
Koh Xuan Yi, Jessica Leong,
Toh Ming Hui and Wang Chen Fong
Structural engineers
TEP Consultants Pte Ltd.
Main contractors PLC Pte Ltd.

Adapting the concept of Origami — the geometric folding of a piece of paper to different shapes and forms — we attempt to depict the parent-child relationship in the way their dwelling spaces are intricately juxtaposed within the sculpted volumes created by the folding of bronze metal planes. While the child is being given her own private space, the parents still want to be able to watch over her from time to time and remain a big part of her life. Hence, the junior master suite has a strong visual dialogue with the master suite in the lower living and the upper study while the other remaining areas enjoy full privacy from the parents' supervision. The bronze metal "sheet" forms both the roofscape of the house as well as the base of the bedroom volumes.

The sculptural volumes resulting from the Origami-like folds of the roof planes are complemented by several other sculptural elements or planes within the house like the suspended timber staircase cum vanity to basement powder room, the staircase to the basement, the suspended light shaft and the shading trellis over the living room void.

To facilitate maximum interaction within the very small family, the idea of "to see and be seen" is also extended throughout the house. Activities and movement within the house are being "shared" through the various connecting voids, bridges, picture windows and reflections off the reflective black glass surface. Even the swimming pool that takes on the role of a garden mural of bold floral prints is visually connected to the basement and hence lends a pictorial backdrop to the various rooms in the basement.

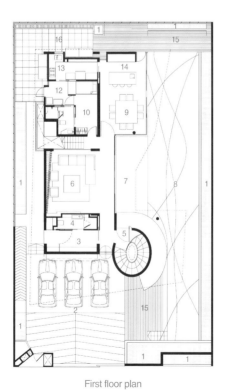

First floor plan

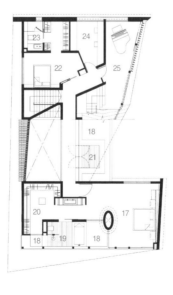

Second floor plan

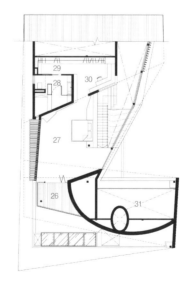

Third floor plan

1. Planter
2. Carporch
3. Shoe Closet
4. Powder Room
5. Main Entrance
6. Living
7. Outdoor Terrace
8. Swimming Pool
9. Dining
10. Maid's Room
11. Maid's Bath
12. Laundry Room
13. Wet Kitchen
14. Dry Kitchen
15. Pool Deck
16. Utility Yard
17. Master Bedroom
18. Planter
19. Master Bathroom
20. Walk-in Wardrobe
21. Outdoor Terrace
22. Guest Room
23. Guest Bath
24. Study
25. Junior Master Lower Living
26. Roof Terrace
27. Junior Master Suite
28. Bathroom
29. Walk-in Wardrobe
30. Study
31. Void Over Master
32. Skylight Void over Basement

Siglap路97号住宅 Ninety 7@ Siglap Road

Location Siglap Road, Singapore
Site area 842.29㎡
Floor area 796.37㎡
Architect design AAMER
ARCHITECTS
Designer Aamer Taher
Photographer Patrick Bingham – Hall

Magnificent views towards the city from atop Siglap Hill inspired this design. Master and family rooms are placed on the third level having the best views.

Living and dining rooms are placed on the second level, connected with external verandas/terraces that flow upwards and fold into the roof form with deep overhangs for sufficient shade, channelling the breeze through the whole house.

Two resort style "cabana" bedrooms are located on the ground level by the pool, with a large open/covered terrace for poolside parties. A sculptural metal "drum" anchors the "ship" to the ground and houses the toilet/shower and barbecue pantry. Roof gardens and timber decks provide added insulation from the sun.

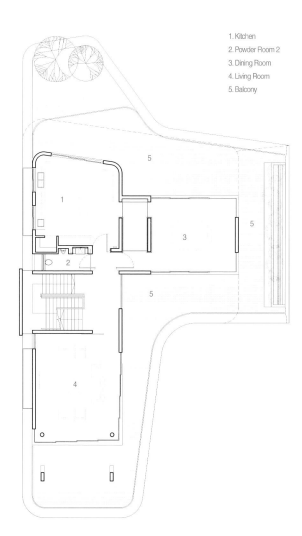

1. Kitchen
2. Powder Room 2
3. Dining Room
4. Living Room
5. Balcony

Second floor plan

1. Car Porch
2. Entrance Foyer
3. Shoe Room
4. Bedroom 1
5. Bedroom 2
6. Laundry Room
7. Household Shelter
8. Maid's Room
9. Powder Room 1
10. Outdoor Deck
11. Deck Seating
12. Cabana

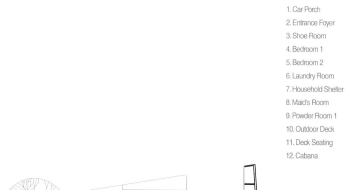

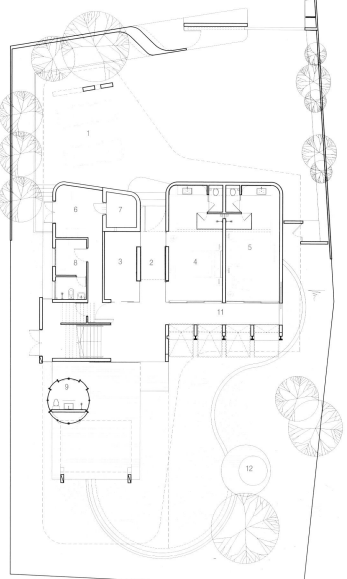

First floor plan

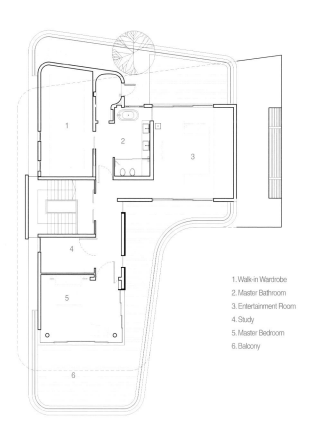

1. Walk-in Wardrobe
2. Master Bathroom
3. Entertainment Room
4. Study
5. Master Bedroom
6. Balcony

Third floor plan

联排别墅 Terrace Villa

Location Baoshan Town, Hsinzhu,
Tiwan, China
Site area 753㎡
Built area 308㎡
Gross floor area 549㎡
Architect design Solid Space Atelier
Design participation Yu-chia Shih,
Designer Su Yon, Co-designer/
Interior designer
Photography Su Yon

Located at the foothill of an enclosed two acre wooded site, the housing design addresses the synthesis of two potentially opposing conditions: the modernistic model of a house with traditional Chinese building typology known as "四合院" or Chinese quadrangles as genealogy by rethinking later model not as discrete entity but a formal and organizational strategy. This specific site also inspires us to employ topological models which operate at two scales: a volumetric organization which allows continuity between landscape and building and a fine scale surface striation that both integrates and articulates geometry and material as they shift from the intensive space of the interior to the extensive space of the exterior.

Three light wells that symbolize the main programmatic intervention were nested within the modules established by the structural grid, and generated private internalized outdoor spaces geometrically driven by the overall organization. On the external expression, the set back volumes promote a more didactical relationship between building and landscape, achieving a certain level of privacy and comfort.

Sectional qualities of these terraced volumes with light wells encourage natural ventilation within the units and floors as well as creating a space in displaying the qualities of light and shadow throughout the day.

Sketch

Model

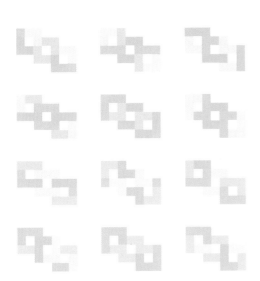

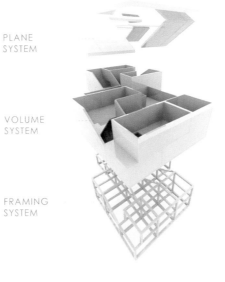

PLANE
SYSTEM

VOLUME
SYSTEM

FRAMING
SYSTEM

Diagram

Diagram

Diagram

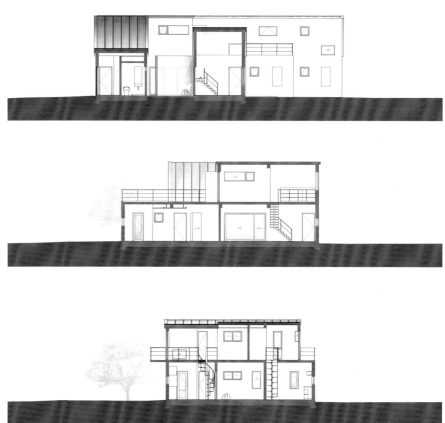

Section

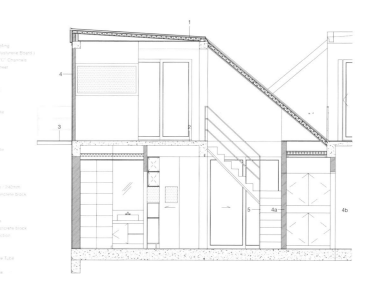

Details

1. Roof Details:
    1.5mm galvanized steel roofing
    120mm XPS (Extruded Polystyrene Board)
    within 200mm roll formed "C" Channels
    0.5mm Galvanized Steel Sheet
    Waterproofing Membrane
    13mm Fiber Plasterboard

2. Second Floor Details:
    15mm Laminate Flooring
    30mm Mortar Bed
    150mm Reinforced Concrete
    10mm Internal Plaster

3. Terrace Structure:
    15mm Tiles
    30mm Mortar Bed
    120mm Reinforced Concrete
    25mm External Plaster

4. Wall Sections:
    External:
    20mm External Plaster
    624mm x 300mm x 120mm / 240mm
    Autoclaved Lightweight Concrete block
    15mm Wall Plaster
    Internal:
    a. 15mm Plaster
    624mm x 300mm x 120mm
    Autoclaved Lightweight Concrete block
    b. Typical Drywall construction
    12.5mm Plaster board

5. Stair Detail:
    50mm x 25mm Stair Square Tube
    10mm Steel Rod
    Customized Size Steel Plate

Detail section

1. Light Well
2. Bedroom
3. Bathroom
4. Restroom
5. Kitchen
6. Laundry Room
7. Living Room/Kitchen
8. Double Height Space
9. Entry
10. Terrace
11. Rooftop Terrace
12. Studio/Study Room

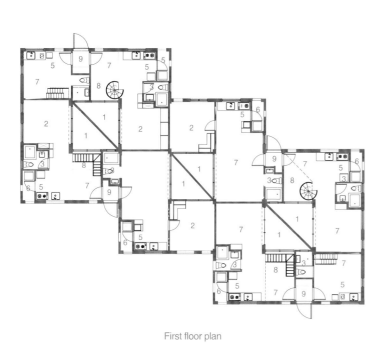

First floor plan

Second floor plan

SK住宅 SK Residence

Location South Jakarta, Indonesia
Site area 800㎡
Architecture design Atelier Cosmas
Gozali
(PT. Arya Cipta Graha)
Designer Arch. Dipl. Ing. Cosmas D.
Gozali, IAI (Principal Architect)
Photographer Tectography
(Fernando Gomulya)

The distinctiveness of this building is that two conceptual ideas are integrated into a single structure here, that is, an urban-resort. The front section of the building (north) adopts an urban architectural style with a dynamic interplay of geometric planes, while the interior section (south) proposes the nuances of a resort. The spatial layout is arranged so that the inner and outer spaces are unified. Resort nuances are also present in the middle, as well as in the back of the building, in the form of an inner court surrounded by three building masses. The inner court is designed from the ground floor up to the top to serve as a center of orientation for the interior space. As the house is located in a water-catchment area, it occupies no more than 30% of the 800 m^2 of land available. What the interior concept aims to bring out is a "clean look interior" dominated by white wrapping in nearly every part of the house. Yet the use of zebrano-motif, high-pressure laminates (HPL) and bright colors like the red and blue in the children's bedroom, makes the space more cheerful and dynamic, not pale.

1. Carport
2. Station
3. Generator
4. Rest
5. Front Terrace
6. Garage
7. Foyers
8. R.Sharlat
9. Study Room
10. Restroom
11. Garden
12. Kitchen
13. Sub Kitchen
14. Terrace
15. Dining Room
16. Family Room
17. Guest
18. Storage
19. Back Terrace
20. Swimming Pool

21. Tempat Jemur
22. Laundry
23. Work Room
24. Maid Bedroom
25. KM/WC
26. Child Bedroom
27. Family Room
28. Balcony
29. Main Bedroom
30. Foyer
31. Wardrobe
32. Guest KM/WC
33. Void

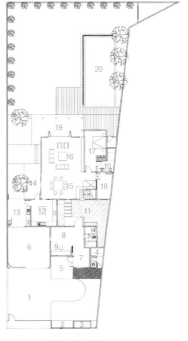

First floor plan

Second floor plan

Alam家庭住宅 Alam Family Residence

Location Jakarta, Indonesia
Site area 750㎡
Built area 1,090㎡
Architect/Interior id-ea (www.id-ea.com)
Structural, MEP, Contractor, Millwork Arsitek dan Rekan Sehati
Photographer Fernando Gomulya

a Private Residence Fostering the Heart of Interactive Family Lifestyle

A residence to host two generations and three households in a single three-storey building.
The different levels — one for the parents and two separate ones for the son's family and the daughter — are designed to balance a certain amount of privacy with a level of interaction.

The frontage of the house is in a way antithetical to the surrounding neighborhood, in which houses are mostly covered with massive fences and guard booths. Alam Family Residence prefers sharing the front yard through a nearly-transparent fence and pushes the privacy screen inward onto the facade, embracing new found relationships between indoor / outdoor and public / private.

The highly articulated concrete west-facing wall with rational modular perforations acts as a graphical breathing brise-soleil, preventing overheating on the building skin, filtering the abstract light qualities and transforming the space throughout the day and night, while delivering a new appearance, indeed a new identity for the twenty-first century residential world.

The interior spaces behind the "graphical breathing brie-soleil" are crystalline white, providing a blank canvas for the constantly

changing visual texture interplay of light and shadow.

The interior of the house is a series of free-flowing, continuous spaces which are organized around the high volume inner voids that bring light and air deep into the house through both plan and section, while fostering a supportive, interactive family lifestyle. The consistent minimal white palate in the common area gives visual dominance to the bold red prayer niche — representing the family's tradition in a modern way, the bold yellow aquarium located in the heart of the dining-living areas, the dark wood "rolling carpet" of the staircase, the combination of marble and terrazzo dining floor which define the high volume space and the vegetation in the courtyard.

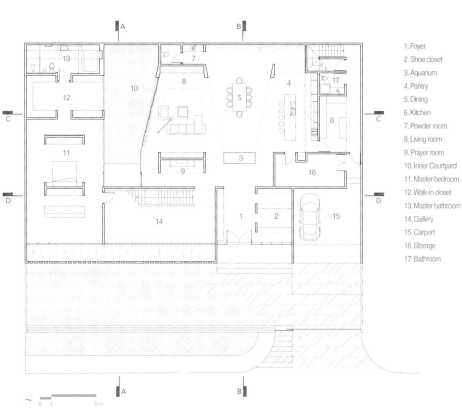

1. Foyer
2. Shoe closet
3. Aquarium
4. Pantry
5. Dining
6. Kitchen
7. Powder room
8. Living room
9. Prayer room
10. Inner Courtyard
11. Master bedroom
12. Walk-in closet
13. Master bathroom
14. Gallery
15. Carport
16. Storage
17. Bathroom

Ground floor plan

1. Gallery below
2. Family room
3. Home theater
4. Open to below
5. Courtyard below
6. Pantry
7. Bedroom
8. Bathroom
9. Balcony
10. Dining below
11. Guest bathroom
12. Guest bedroom
13. Master bedroom
14. Reading room
15. Walk-in closet
16. Master bathroom
17. Roof deck
18. Roof garden
19. Mechanical roof
20. Skylight
21. Service
22. Washing & Ironing
23. Maid room

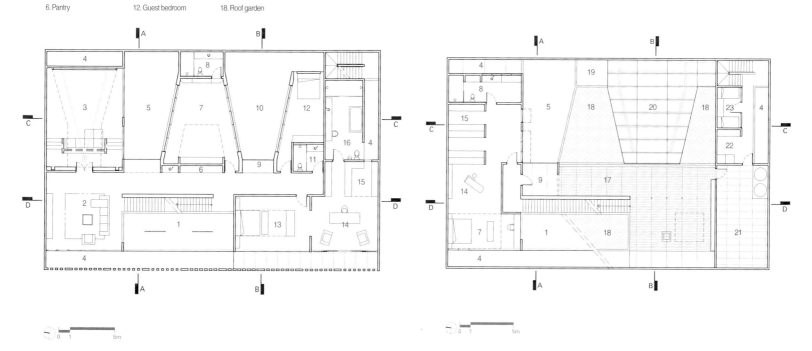

Second floor plan

Third floor plan

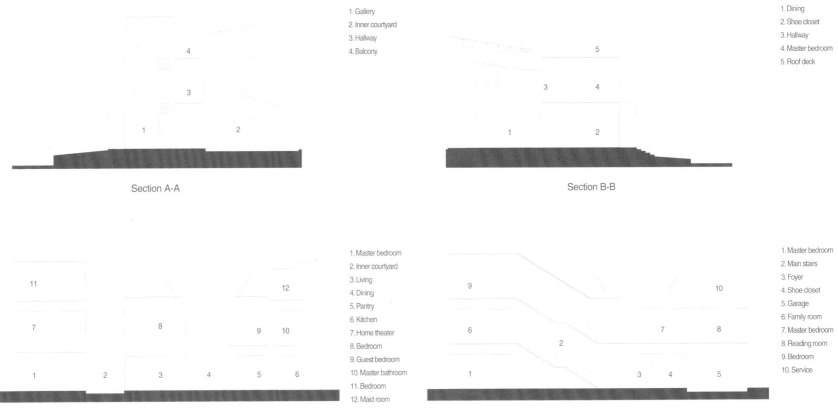

1. Gallery
2. Inner courtyard
3. Hallway
4. Balcony

1. Dining
2. Shoe closet
3. Hallway
4. Master bedroom
5. Roof deck

Section A-A

Section B-B

1. Master bedroom
2. Inner courtyard
3. Living
4. Dining
5. Pantry
6. Kitchen
7. Home theater
8. Bedroom
9. Guest bedroom
10. Master bathroom
11. Bedroom
12. Maid room

1. Master bedroom
2. Main stairs
3. Foyer
4. Shoe closet
5. Garage
6. Family room
7. Master bedroom
8. Reading room
9. Bedroom
10. Service

Section C-C

Section D-D

0 1   5m

Soo Hwa Rim住宅 **Soo Hwa Rim**

Location Chungcheongnam-do,
Korea
Site area 976㎡
Bldg. area 266㎡
Total floor area 371㎡
Architecture Design Design Group
OZ/ Lim Sang-jin, Shin Seung-soo
Photographer Lee Jeonghun,
Lee San-muk

Hide and Seek

The site of this project, Seosan area & vicinity belongs to Taean peninsula, north-western part of Chungnam, where was a geographical outpost for import trade connecting with China, just as Seosan MAESAMZON Buddhist Image is seen differently by light angle, the place is also well-known for cultural region from which Asca ancestor art affected and breathe of people of Bakjae state is still alive. The program that was asked to design as an architecture's living space & guest house for visitors is located in the place to see a beautiful scenery like Mt. Gaya and vicinity lake, in which there are 4 houses and single house with independent 6 programs to show off its own being. The designer has a point to differentiate the feature of the spot in the middle of changing atmosphere to give a common landmark without losing its uniqueness.

The important feature of architecture element of "Soo Hwa Rim" is to embrace surrounding mountains & large ground to bring the entire space like lake & surrounding scenes to make rooms or balconies that split them respectively. Balconies infiltrated into whole volume in various way is countermeasure equipment to respond changing four seasons, it changes by itself over times, just like standing Seosan MAESAMZON Buddhist Image, and it reflect its antique scene to breathe with residents connecting & disconnecting inside from outside. Whereas upright louver, final material that applied on external wall exist as surface layer that connects spaces giving different shapes by perspectives. On this place that seems like a spur, everyone can experience hiding-finding game.

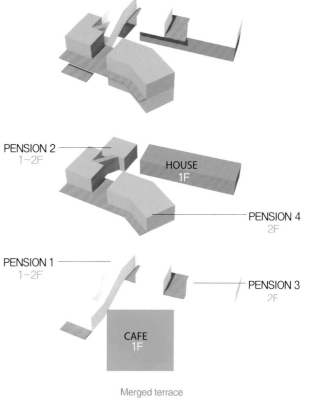

PENSION 2
1~2F

HOUSE
1F

PENSION 4
2F

PENSION 1
1~2F

PENSION 3
2F

CAFE
1F

Merged terrace

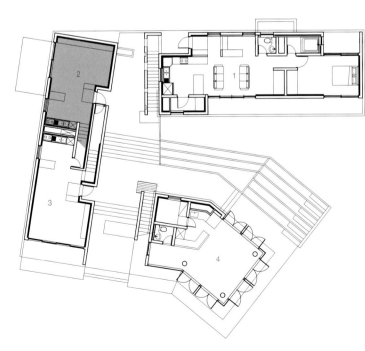

First floor plan

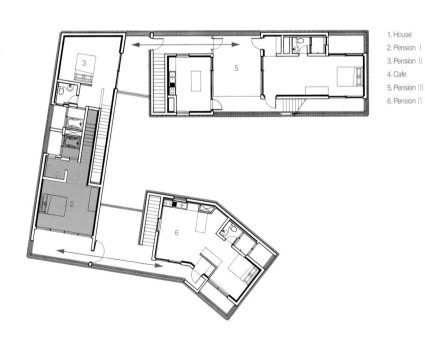

Second floor plan

1. House
2. Pension I
3. Pension II
4. Cafe
5. Pension III
6. Pension IV

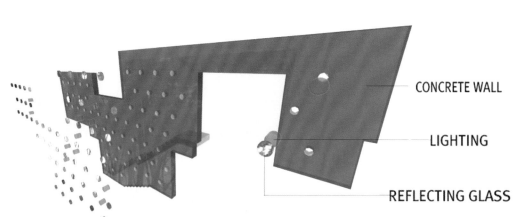

CONCRETE WALL

LIGHTING

REFLECTING GLASS

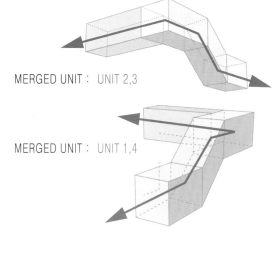

MERGED UNIT : UNIT 2,3

MERGED UNIT : UNIT 1,4

CONSTELLATION

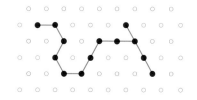

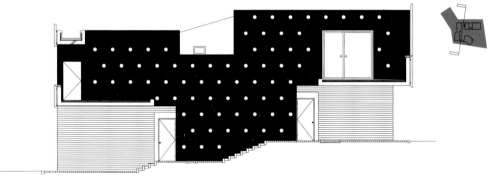

Section I

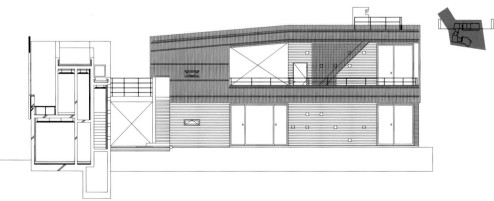

Section II

Mook Heon住宅 **Mook Heon**

Location Gyeonggi-do, Korea
Site area 215㎡
Blgd. area 81㎡
Total floor plan 254㎡
Architecture design Koossino /
Koo Seung-min

The Mook-heon receives the diving of light and wind in a natural way. An underground space is made along the courtyard that vertically penetrates the heart of the building. A vertical path covering all floors is linked to the underground space. The ceiling transfers sunlight down to the underground space by absorbing the light source in its flat space. The volume of the land had been returned to the underground space so that the width and depth of the rich space can be secured and taken out. The Mook-heon boldly took out the above ground volume while having the underground space in the center of everyday living. The added space is expanded as a yard space. The upper part along the long axis of the road was built in cantilever structure. If the underground floor is the space for family gathering of

the owner, the small space on the ground floor supporting the upper floor is the center of emotional mentality. The inconvenient vertical movement-line that links the underground and ground floor along the courtyard is located to face south. It is planned as a medium space that transfers the projected light to the opposite courtyard. The client had asked to float the privacy of Mook-heon up in the air so that it would not be interfered by the ground. Accordingly, the floating space of the cantilever is located along the track of everyday living that looks like a long drape of a mass. The road opens the spaces that cross each other in all directions to extend the flow of the yard, which is also the boundary of the land. Because of it, the mass was kept at the column-less space of cantilever. The angulated

mass which is pressed long added of agile acute angle had been camouflaged along the long axis of the land. The camouflage of the heavy weight is the transparency of the Marblelite. The diffusion of the reflected light had been expected considering the property of exterior finishing material. It had been induced to the flow toward north so that only the calm silhouette would be contained. The Mook-heon is a displacement entity that cannot live by itself in the residential compound; however, a new genetic interpretation of residence is expected by placing a mutant entity.

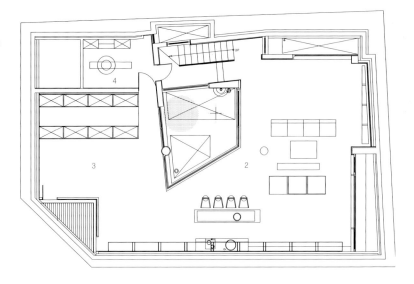

1. Entrance
2. Studio
3. Hall
4. Boiler room
5. Corridor
6. Room
7. Toilet
8. Terrace

Basement floor plan

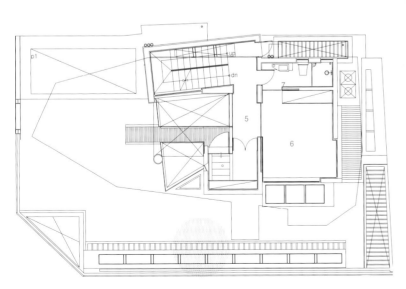

First floor plan

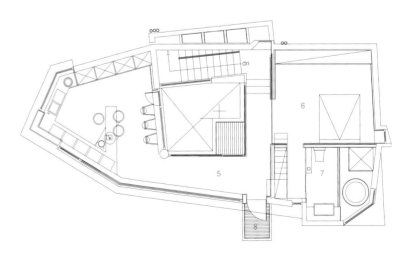

Second floor plan

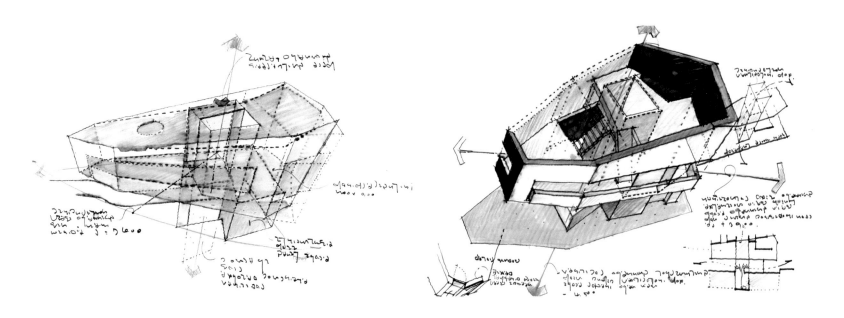

Sketch

蚯蚓住宅 Earthworm House

Location Gyeonggi-do, Korea
Site area 628㎡
Blgd. area 192㎡
Total floor area 326㎡
Architecture design NODE
ARCHITECTURE DESIGN/ Kim Won-ki
Design participation Ryu Kyung-jin,
Ko Seong-eun
Interior design Shin Hyun-jeong,
Kim Yoon-sook

The project site has a good river view. It is located on the steep northern slope at the entrance of Green Forest Community Village, which in front of the Green Forest Waldorf School. The design visualized the shape of two earthworms in the process of mating. A large earthworm made of yellow marble stone in the upper part and the front is holding up its head toward the pond. The small earthworm made of high-density red wood panel has half of its body in the ground. The two earthworms are making the space and the gap between them by way of their movements. The floor over the bridge is planned as a space that changes in accordance with the variation of games, family members and season. The nest of the earthworm gets the energy for air-conditioning from the subterranean heat coming

from underground 130 m depth. It also gets its electricity from the solar energy panel. The pilotis space under the living room next to the pond and the roof of the right side guest room are public spaces that can be used by the Green Forest School Community Village. The positive role of the earthworms is anticipated in this respect.

West elevation

North elevation

1. Workshop
2. Yard
3. Guest room
4. Parking lot
5. Living room
6. Kitchen
7. Corridor

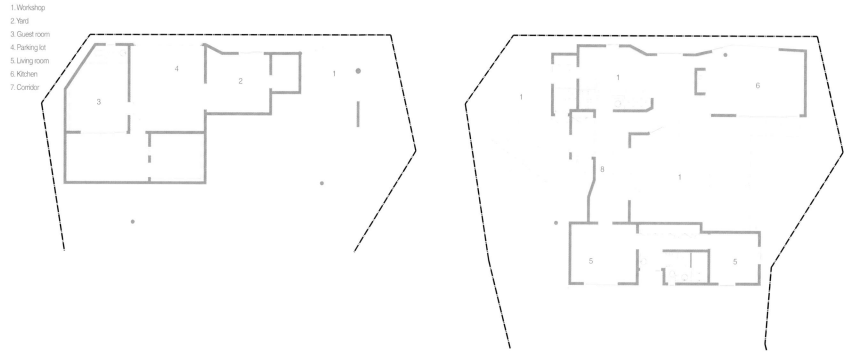

Basement floor plan   First floor plan

1. Room
2. Corridor
3. Deck

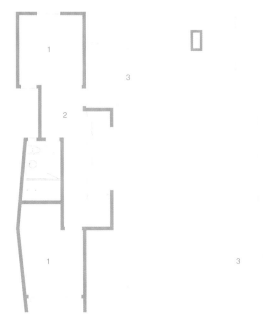

Second floor plan

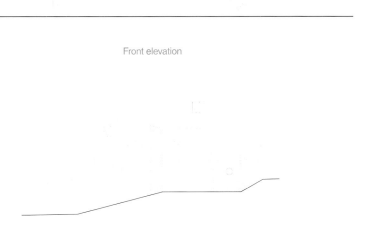

Front elevation

Rear elevation

Ee-eep-jae住宅 **Ee-eep-jae**

Location Busan, Korea
Site area 743㎡
Bldg. area 274㎡
Total floor area 387㎡
Architecture design Toban
architecture associates / Kim
Jeong-gwan
Desigd participation Jang Hyun-uk,
Cho Mun-ki, Hwang Chan-uk

"Ee-eep" is the two paths which are in harmony with each other in the arriving at the truth. One is the "rationality", which is the path of reason with high-class qualification in meeting the world. The other is the "practice", which is being comfortable at home having returned from the outside world.

I interpreted the "Ee-eep-sa-haeng-ron": two paths and four methods in reaching the truth theory of Dharma in the design of this house.

There are two divided spaces. One is the square next to the road and the other is the hidden courtyard. The square has three pine trees and a waterscape entering the porch and the front door. It is an external spaces with a class as the "rationality". The courtyard, facing the dining room and the main room, is

the space of the "practice", making the internal and external spaces in one, both visually and spatially.

The internal space of building is divided into public space and private space. The public space is in the entrance area. The pilotis makes the parking lot while lifting the living room like a gazebo. It expresses the "rationality" by being a visual main view target of the house seen both from outside and from the square.

The private space is in the mezzanine location. It binds the lives of family members into one by linking the building having rooms. The rooms in the private space are the "practice" being the stationary space located in the internal of the overall space.

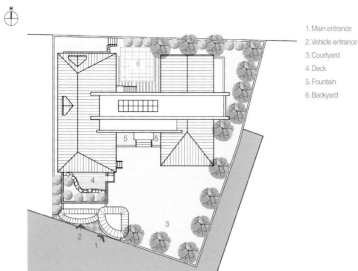

1. Main entrance
2. Vehicle entrance
3. Courtyard
4. Deck
5. Fountain
6. Backyard

Site plan

1. Parking lot
2. Entrance
3. Storage
4. Office
5. Fountain
6. Study room
7. Dressing room
8. Master bedroom
9. Bedroom 1
10. Bedroom 2
11. Backyard
12. Family room
13. Roof
14. Balcony
15. Utility room
16. Kitchen / Dining Room
17. Living room
18. Deck

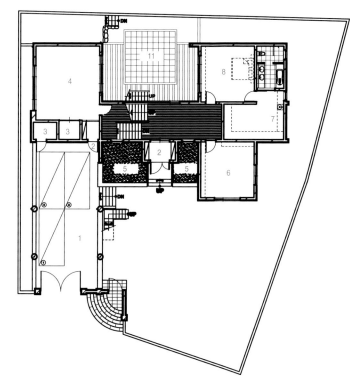

First floor plan

Second floor plan

1. Entrance
2. Corridor
3. Family room

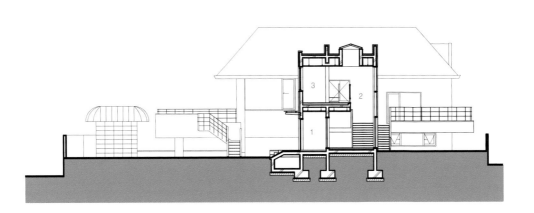

Section Ⅰ

1. Entrance
2. Storage
3. Corridor
4. Study room
5. Bedroom
6. Family room
7. Living room
8. Attic

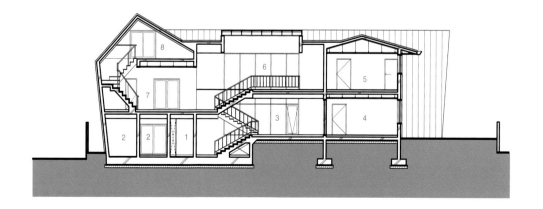

Section II

"漂浮" 住宅 Floating House

Location Geonggi-do, Korea
Function Residential facility
Bldg. area 290㎡
Architects Hyunjoon Yoo
Architects

The house was located in the north as far as it can be for a big sunny yard in the south. The guest house was located in the south so that the restaurants and motels are not seen from the yard. The guest house was built as if it was the fence to be located as far as it can be in the south. It looks like it was put inside the thick fence. To view the river from the yard and the guest house, the main building was leveled up to be a piloti. The roof of the main building was made as a plain roof and there is a roof garden in which people can overlook the river from high level. As a result, the river is viewed from every part of the house.

The plan can be described as the "space within space". The service space, such as a bathroom, a dress room, and a kitchen, was concentrated in the middle and compactly distributed, and a bedroom was placed in the east, a living room in the west, and a corridor in the outskirts. To make people in the house feel the space bigger than real, a circular traffic line was adopted instead of one main traffic line. The circular traffic line was made with a minimum width to differentiate the intensity of space, but the space does not look small because the traffic line extends to visually meet the terrace. Since the stairs that lead to the roof were made as if they were inserted to the second floor, the staircase does not protrude on the roof. The staircase itself plays a role as a well of light, transmitting light to the middle level of the house.

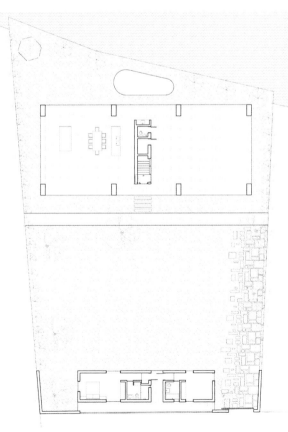

Ground floor plan

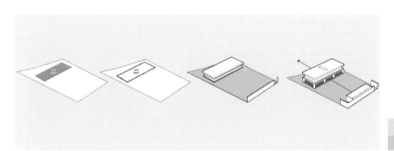

Arrangement plan

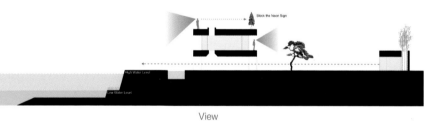

View

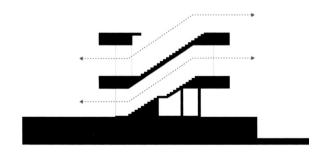

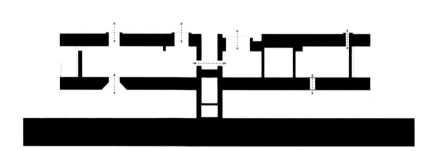

Circulation plan

1. Living room
2. Kitchen
3. Bathroom
4. Library
5. Dress room
6. Master bedroom

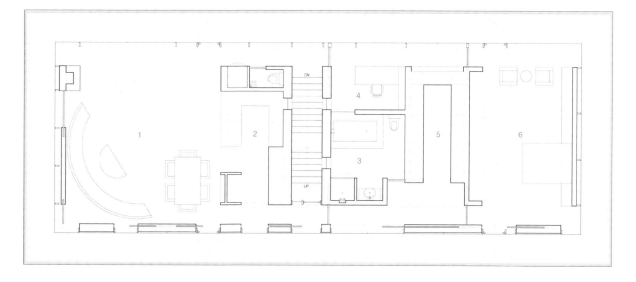

First floor plan

Acrylic 住宅 Acrylic House

Location Yamanashi, Japan
Site area 1,024㎡
Bldg. area 73㎡
Architecture design Takeshi Hosaka
Architects / Takeshi Hosaka
Design participation Takeshi Hosaka
Photographer Toshihiro Sobajima

This building is a residence for a young husband and wife and their three children. It stands amid rice fields and houses, located a little off the main road for sightseeing. So, we have decided to design the house in such a way that the inside is always in contact with the outside, creating a sense of open space, yet it is hard to see what is going on inside the house from outside the property, making the boundary between the inside and the outside as "transparent" as possible.

To be specific, we have made the extremely transparent, continuous boundaries between the inside and the outside that don't have joints or mullions by using heat-insulating, 20㎜-thick acrylic boards. Acrylic fiber has very few impurities, so the surface doesn't reflect much light even during the day, making

it easy to see the inside from the outside quite transparently. Just as you can see the garden from the living room, kids playing in the garden can see the inside, too.

Each side has an opening for entering and exiting, making it easy to access the garden and the terrace from anywhere inside the house. The kids can enjoy both the inside and the garden as a place for playing and other activities, while the adults can go out to the garden and hang laundry, or they can also enjoy meals and tea as a family. Usually, life inside the house and life in the garden are each one continuous experience, but now that the inside and the garden are divided only by the extremely transparent, continuous boundaries with no joints; it feels like they are more intermingled than before.

1. Garden
2. Dining area
3. Bed room
4. Terrace
5. Children's room
6. Bunk beds

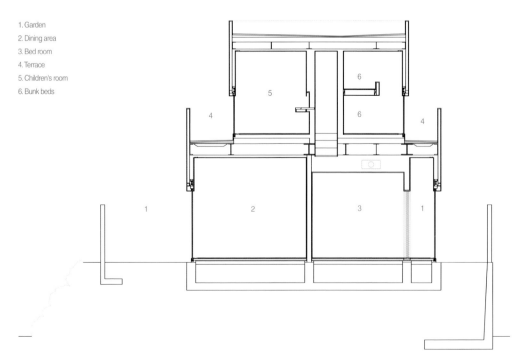

Section

1. Garden
2. Entrance
3. Living area
4. Dining area
5. Kitchen
6. Bed room
7. Corridor
8. Bath room
9. Powder room
10. Terrace
11. Children's room
12. Bunk beds
13. Storage

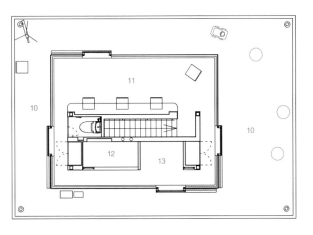

Second floor plan

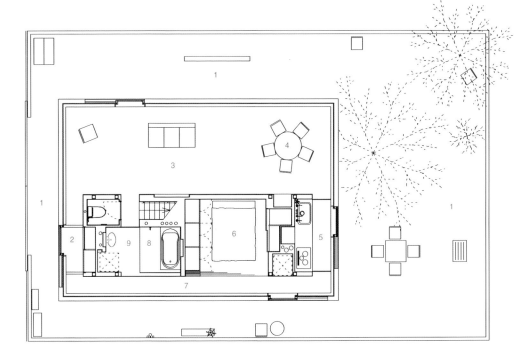

First floor plan

"绿色明天" 住宅楼 Green Tomorrow

Location Gyeonggi-do, Korea
Site area 2,456㎡
Bldg. area 622㎡
Total floor area 715㎡
Architecture design Samoo Architects &
 Engineers / Kim Kwan-joong
Photographer Yum Seung-hoon

"Green Tomorrow" adopted environment-friendly approach utilizing passive design from the beginning of planning. It also introduced the actual live-in possibility and the concept of Korean tradition. Project site is a long shape on east to west axis. The design considered land and environmental aspect, such as the view toward the Zelkova tee adjacent to the site. The spatial planning of the two buildings expressed the integration as a whole and the individuality as each building together with outdoor space. For this, the integrity with outdoor space is maintained having the garden and outdoor corridor in the center. Each building has its outdoor space so that it will have both integrated space and individual space. The building faces south and it is located long from east to west. It minimizes the

energy consumption and maximizes the use of solar heat and sunlight together with rooftop landscaping. In the unit plan, the indoor space and outdoor space are interacting with each other through the technical elements and emotional elements. Environment-friendly materials made of natural raw materials are chosen for major exterior finishing materials. The roof shape considered both rooftop landscaping and solar heat power generation through energy simulation. The flat portion of roof has vegetation and sedum vegetation has been applied. It makes the image of the building softer, which might have been too dry caused by conspicuous solar heat power generation facility. It also lets the building escape from the simple shape of single story building. If a general shape roof had been used,

when the south side of the roof would be used for PV panel, the north side cannot be utilized in full. North side can have landscaping; however, vegetation would be limited and it would be sloped vegetation. We designed a shape which has both flat and sloped and solved this difficulty.

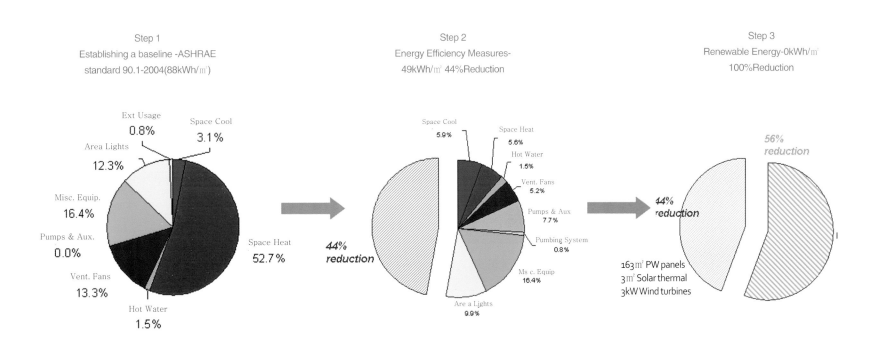

Step 1
Establishing a baseline -ASHRAE
standard 90.1-2004(88kWh/㎡)

Ext Usage
0.8%
Space Cool
3.1%
Area Lights
12.3%
Misc. Equip.
16.4%
Pumps & Aux.
0.0%
Vent. Fans
13.3%
Hot Water
1.5%
Space Heat
52.7%

44%
reduction

Step 2
Energy Efficiency Measures-
49kWh/㎡ 44%Reduction

Space Cool
5.9%
Space Heat
5.6%
Hot Water
1.5%
Vent. Fans
5.2%
Pumps & Aux
7.7%
Pumbing System
0.8%
Ms c. Equip
16.4%
Are a Lights
9.9%

44%
reduction

Step 3
Renewable Energy-0kWh/㎡
100%Reduction

56%
reduction

44%
reduction

163㎡ PW panels
3㎡ Solar thermal
3kW Wind turbines

Energy Consumption

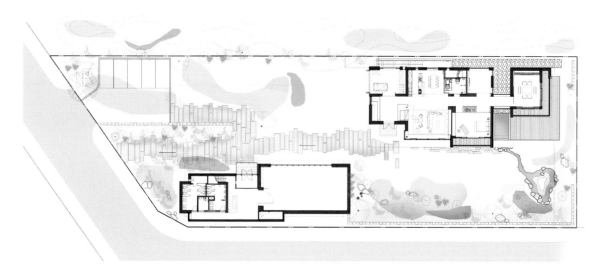

Floor plan

橘色住宅 Orange House

Location Bilkent, Ankara, Turkey
Total construction area 1,050㎡
Architecture design Yazgan Design Architecture
Photographer Yunus Özkazanç 2010, Kerem Yazgan 2010
Employer Isabel Knauf, Peter Redecke

Orange House is a 1050m² residence of three storeys, located in a very steep site. The site has a nice combination of urban view with Middle East Technical University forest. The building has a steel structure. The main principle that shaped the design process is the program written by the architects by considering the one given by the employers. The program is based on the development of flexible relationships between diverse inputs of design, such as materials, program elements, demands of users, dimensional requirements, site peculiarities, Ankara climate and its habitat, architects and engineers involved with the project. Writing the design program can be called as writing the "design of relationships". The "design of relationships" refers to the relations developed between not only design programs, but also between the drawing and the architect, the architect and the owner and the owner and the drawing. The design program is supported by diagrams, giving way to the integration of the users with the design process. The building is the product of a flexible-systematic process without losing the initial idea. The building design is based on surfaces that develop an exterior-interior space relation between autonomous rooms, and an initial diagram concerning the circulation connecting these rooms. All rooms reflect the specific needs of employers in their design. For instance, dining room is dimensioned with reference to the existing Persia carpet belonging to the users. Every room wall is double-layered. They took into shape by taking advantage of a

peculiarity brought forward by steel construction. Shafts, niches, doors, rainwater pipes, wardrobes, etc., are located at the inside surfaces of the double layers. Through that approach, technical and user requirements are integrated with the initial diagram. The steel structure follows a 60cmx60cm grid dimensioning in the design, therefore, provided another systematic for diagrams.

Elevation

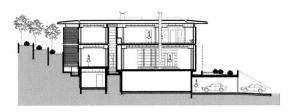

Section

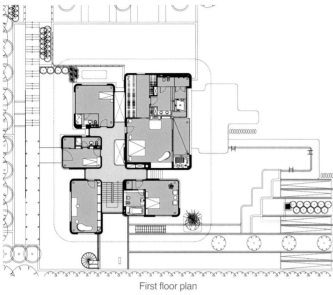

First floor plan

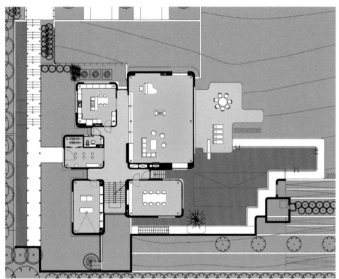

Ground floor plan

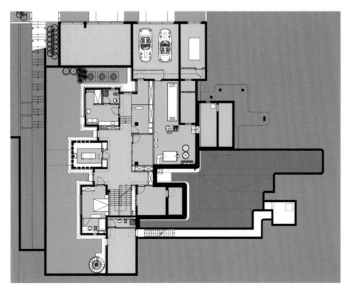

Basement floor plan

YSY住宅 YSY HAUS(Wear House)

Location Seto, Japan
Site area 183㎡
Bldg. area 67㎡
Total floor area 92㎡
Architecture design AUAU /
Akitoshi Ukai
Photographer Yoshimura Masaya

Outline of design

It is a small house for two small children and a young couple. The blank and coming in succession were invented by making the best use of the vertical interval in the small hill. This creates an inside, outside new relation and a sense of distance, reinforces the subject soft "Enclosure" structurally by "Wear", assumes the strengthened plan, and has the role to ease the thermal environment in the environment.

Individuality that makes the best use of piece

I walked with the client and to the town, searched for land. Surrounding town was collected; the planning was decided by the process of thinking about the family's life together, and

the cooperation of labor work with the client was proceeded. It thought about the cooperation arranging the private room on the first floor, and the living space that held the view concurrently was distributed to the second floor. The ideal way to which the piece or more helped by esteeming the individuality of each life, and bringing it together by another individuality was requested. The second floor becomes openhearted at a dash, and actually feels the existence of "Wear" while the first floor is a set of the box when entering an internal space. The scenery seen from the window can be intentionally experienced as turned over.

Structure

This building's two wooden stories. It aimed at individuality as the whole and the acquisition of strength by treating wooden by bundling, that is, "Enclosure" and "Wear" two this time, and treating the relation between them. The spreading wooden frame that consists of basic vertical and the horizontal brace is adopted, and "Enclosure" part has been passed by a minimum structure. On the other hand, "Wear" part composes of the diagonal material the streak or exist, and strengthens "Enclosure" that becomes a life scene.

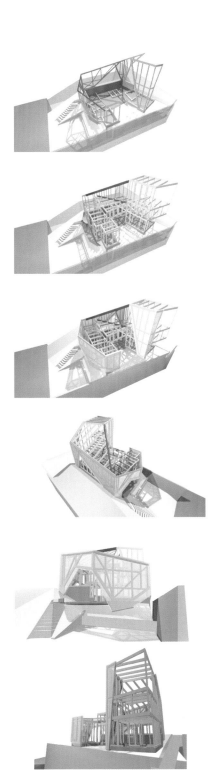

3d modeling(structure)

1. Entrance
2. Studio
3. Bath room
4. Play room
5. Children's room
6. Bed room
7. Garden
8. Dining room & Kitchen
9. Living room
10. Loft

Roof floor plan

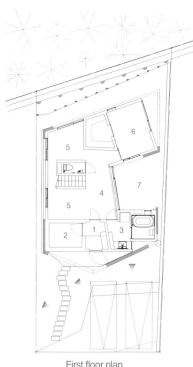

First floor plan

Second floor plan

North elevation

South elevation

East elevation

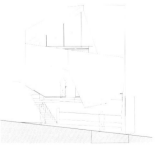

West elevation

内外住宅 Inside & Outside House

Location Tokyo, Japan
Site area 96㎡
Bldg. area 37㎡
Architecture design Takeshi hosaka
Photographer Masao nishikawa

This is a residence consisting of inside house and outside house. It is also a proposal of architecture of a new relationship between internal and external. Because the site is an irregular shape, the planar shapes of the rooms in the two cubes of "inside house" and "outside house" are irregular shapes. In the two cubes, there are glass windows and wooden-board windows in the peripheries, and any room lets in light and wind from the windows and the inside house intermingles with the outside house through the windows. When the wooden board windows are opened, liberating level increases and light and wind further intermingle with the surrounding environment. When the wooden board door at the entrance is opened, it is planned so that the view can run through from the north garden

to the south garden and other wooden board windows are also planned so that the view can run through unexpectedly. The "inside house" houses rooms where a family of four people can live in comfort, places equipped with water supply and electric appliances, etc. It consists of entrance, living room, dining room, TV, air conditioner, kitchen, bedrooms, children's room and bathroom, etc. The "outside house" houses lives that you want to do in the outside house than the inside house. For example, keywords such as woodwork, work, growing plants, keeping insects, maintenance of bicycles, hammock, sunbathing, gardening, playing with water, seeing the sky, tools for mountain climbing, napping and reading books, etc. are assumed. The opening sections can be opened with different

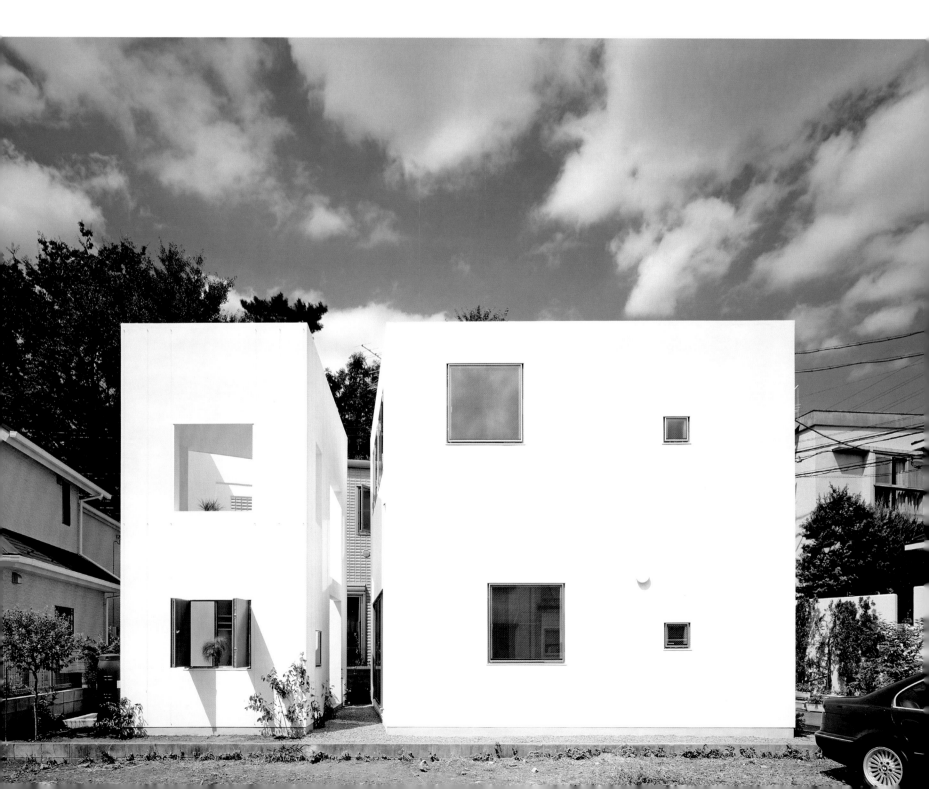

heights and sizes between the inside house and the outside house and they were planned so that you can feel more distant than the actual distance and various places can be generated in the small site.

I thought that many life acts would be richly generated even in a small site in Tokyo by building "inside house" together with "outside house" where you can treat those things such as nature, rain water, soil, sand, animals, insects and birds that were eliminated by the modern architectures and cities, and acts which are originally supposed to be outside house.

Sketch

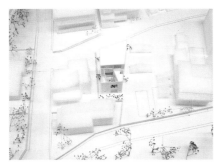

Model

1.Garden
2.Entrance
3.Living room
4.Dining room
5.Kitchen
6.Bicycle maintenance room
7.Outer room
8.Atelier room
9.Room

10.Powder room
11.Bed room
12.Cloth room
13.Terrace
14.Void
15.Flower room
16.Roof terrace
17.Roof

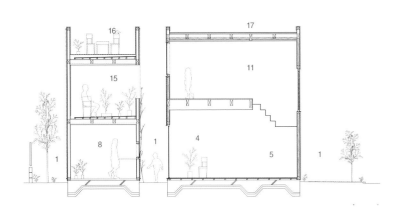

Section

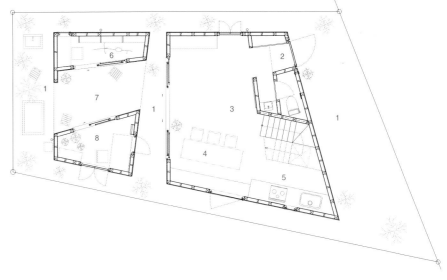

First floor plan

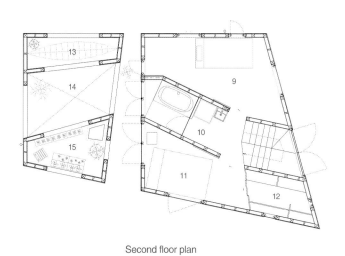

Second floor plan

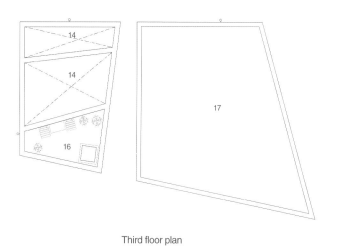

Third floor plan

T住宅 T House

Location Kyoto, Japan
Site area 805㎡
Bldg. area 102㎡
Total floor area 211㎡
Architecture design atelier BORONSKI
 z/ Peter Boronski
Interior design atelier BORONSKI
Photographer Kei Sugino

On a small hill overlooking Kyoto city a suburban house for a young couple negotiates some tough "Historical" design regulations. Many decisions have already been made. Therefore the house is conceived of as a simple container with private spaces lodged randomly within. But the randomness is orchestrated. Despite the external restrictions, the fluidity of the interior void spaces and curved connecting stairways finally let the house feel very free.

There are three primary elements at work in this composition. The main external walls (running east/west), the private volumes (overlapping and bridging) and the resultant void space.

The main Living, Dining, Kitchen area on the first floor opens onto the main terrace facing the garden to the East allowing classical indoor & outdoor living, and the ceiling height in this area varies from 2.5m to 7.5m. The two main bedrooms bridge the building and their North & South facing walls of glass allow the external cladding to continue into the rooms (one white plaster, one black timber). The bathroom on the second floor has a large internal window overlooking the garden to the East and the guest bedroom on the third floor pushes straight out to the West. Beside and below this bedroom are two minor terraces that spatially overlap. The second lounge area on the third floor is just a floor slab, a viewing platform that bridges the main void and allows sweeping views of the city to the East. There are also two top-lights allowing vertical views to the sky.

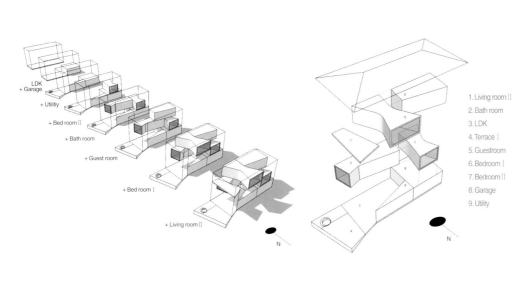

LDK
+ Garage

+ Utilitiy

+ Bed room II

+ Bath room

+ Guest room

+ Bed room I

+ Living room II

N

1. Living room II
2. Bath room
3. LDK
4. Terrace I
5. Guestroom
6. Bedroom I
7. Bedroom II
8. Garage
9. Utility

N

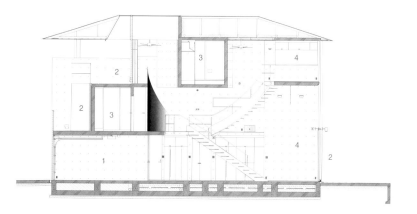

East-west section

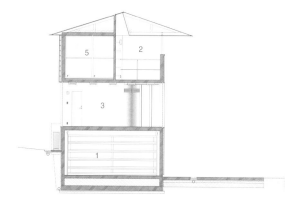

North-south section I

1. Garage
2. Terrace
3. Bedroom
4. Living room
5. Guest room
6. Toilet
7. Kitchen
8. Bathroom
9. Study room
10. Carpet

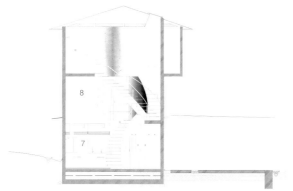

North-south section II

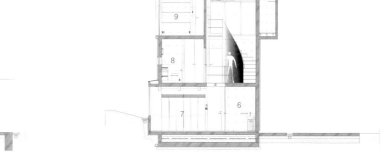

North-south section III

1. Carpet
2. Garage
3. Toilet
4. Kitchen
5. Dining
6. Livingroom Ⅰ
7. Terrace Ⅰ
8. pool
9. SPA
10. Terrace Ⅱ
11. Bedroom Ⅱ
12. Bathroom
13. Terrace Ⅲ
14. Guest room
15. Study room
16. Bedroom Ⅰ
17. Living room Ⅱ

Third floor plan

Second floor plan

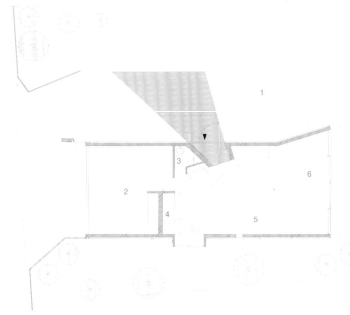

First floor plan

L71住宅 L71 House

Location Bangkok, Thailand
Site area 650 m^2
Architecture design OFFICE AT Co., Ltd.
Surachai Akekapobyotin,
Juthathip Techachumreon,
Pradab Suksamran, Nattakarn Kerdkaew
Interior & Landscape Design
OFFICE AT Co., Ltd.
Photographer Wison Tungthunya

PROGRAM: The L71 house is a single family house located on the northeastern side of Bangkok, Thailand. The site for the house is a long and narrow shape site. All of the house programs such as 4 bedrooms, dining room, and family room are placed along the site to face north direction. Since the owners have some parties occasionally, the public areas, such as living room and parking, are in the front of the house, and the private areas are in the back of the house along with a swimming pool.

EXTEND: The living room mass was extended to create private space for the swimming pool and the second floor mass was extended to create shading for the swimming pool and terrace.

SPLIT: If the house is designed as one big mass it will block ventilation and natural light, so in this house each room is spitted to maximize ventilation and natural light.

INSERT: Since the masses are split, the house creates some semi outdoor spaces to interlock indoor and outdoor spaces. Varieties of natural materials including water, grass and wood are inserted into each space.

COVER: The roof of the main house is double-roof to cover the house from weather. The lower roof is reinforced concrete slab, and the upper roof is metal sheet roof. A space between the two layers of roof acts as an air buffer, natural ventilation and allows for easy maintenance.

MATERIALS: Main material of this house is painted plastered brick wall and tint glass. Where the masses are split, the material of the split masses is wood.

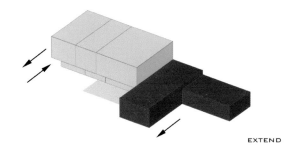

EXTEND

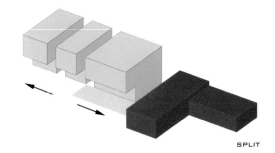

SPLIT

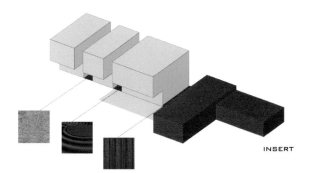

INSERT

Diagram

1. Parking
2. Living Area
3. Dining Area
4. Kitchen
5. Laundry
6. Terrace
7. Swimming Pool
8. Master Bedroom
9. Buddha Room
10. Family Room
11. Working Area

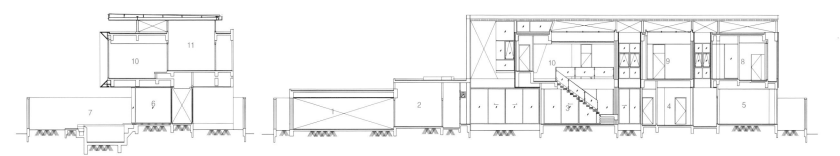

Section A-A

Section B-B

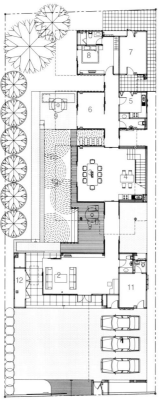

First floor plan

Second floor plan

1. Parking
2. Living Area
3. Dining Area
4. Pantry
5. Kitchen
6. Exercise Area
7. Laundry
8. Bedroom 1
9. Terrace
10. Swimming Pool
11. Maid's Room
12. Storage
13. Master Bedroom
14. Son Bedroom
15. Buddha Room
16. Guest Bedroom
17. Family Room
18. Working Area

盒子住宅 Out of the Box

Location Bangalore, India
Site area 111.5㎡
Building area 214㎡
Architecture design Cadence
Design participation Smaran Mallesh, Narendra Pirgal, Vikram Rajashekar
Photographer Claire Arni

The 111.5m² corner site presented us with the classic urban scenario. The site was abutted by houses on two sides and flanked by low income housing on the other two sides. The question thus posed to us was, "What would be the relationship of the dwelling to the outside?" The stand taken by us was to incorporate the "outside" inside while the building shuns the surroundings. A classic diagram of this would be the traditional courtyard house. Taking this classic diagram we moved the court to the corner to create new spatial and formal effects. By moving the court to the fourth quadrant of the square we could magnify the boundaries of each program flanking the court i.e. the living room, the dining and the bedrooms would not only feel much bigger but also would have sectional relationship with

the open to sky court. The court is further articulated by placing a sculptural element that would serve as an informal dining area as well as a tub for housing a tree. The jali wall cast in-situ completes the fourth corner to accentuate the experience of the court.

This idea of the elevation was to have a customized jali of wall with a pattern of openings which fade away to form the platonic cube. A fiber glass mould of size 2'x2' was made to cast the concrete tiles; the thickness of this tile was 100mm.

The mould had a set of 4 oval openings; these openings were then filled up according to the pattern to achieve the desired

variation.

The tiles were stacked up like a conventional brick wall to construct the wall. These tiles were reinforced with metal flats to ensure stability.

Corner open-to-sky court makes all the spaces with programme open into the same, further extending the boundaries of these spaces. The Garden sculpture is made of brick and cement, which houses a tree and extends to form a table for four.

Thus the building while negating the outside environment simultaneously houses the "outside" inside.

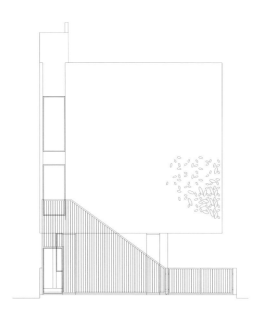

Elevation

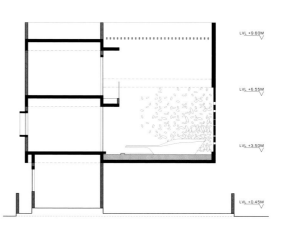

Section

UNFOLDED JALI WALL

PLAN @ COURTYARD LEVEL

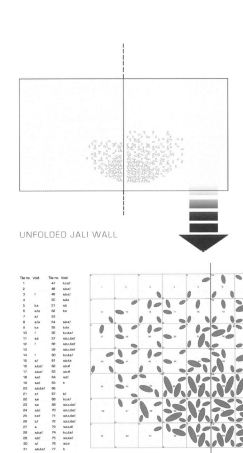

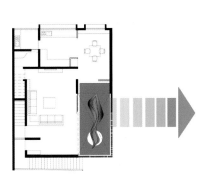

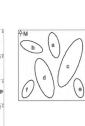

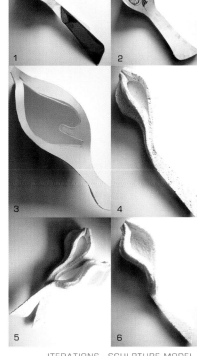

ITERATIONS - SCULPTURE MODEL

-THE COURTYARD SCULPTURE MADE OUT OF CONCRETE WAS DESIGNED TO INCORPORATE AN OUTDOOR DINING AREA AND A TUB TO HOUSE A TREE .

- THE JALI WALL WAS CONSTRUCTED BY MODULAR PRE-FABRICATED CONCRETE TILES. 2' X 2' FIBRE GLASS MOULD WAS USED TO CAST THE CONCRETE TILES

ITERATIONS - FACADE MODEL

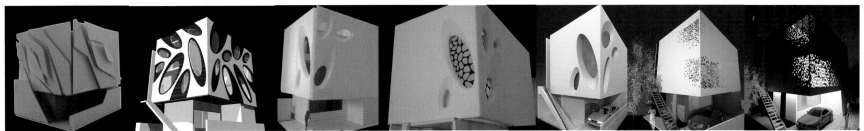

AMA住宅 AMA House

Location Aichi Japan

Site area 464.0㎡

Built area 85.92㎡
Architecture design Katsutoshi Sasaki
+ Associates
Photographer Toshiyuki Yano
Structure company g2plan
Construction company Sunshow
industries ltd

Small House

As the site is surrounded by rice fields, we planned "a small house" that the idyllic atmosphere and landscape.
The entire volume of the house was first divided into individual rooms, their each concept was finally linked together taking account of factors such as connection of garden and room, entrance of light, ventilation, flow line of daily activities, etc.
Also, to meet with the demand for a guest parking lot and family garden, we laid out the rooms across the site to secure two exterior spaces.

Multiple Viewpoint

Each room has different volumes, finishes, and openings.

These differences were made to enhance deeper experience with elements by presenting more than one viewpoint on each element; for example, when the light enters from wide opening, it gives you different impression from the thin ray of light in a dark place.
These elements can be trees in the garden, wind, internal openness, nuance of shadows, and communications between family members.

Connected Air

When opened the door, these rooms become "One single room with connections". Although, unlike a general single room, it can not get a view of whole room, one room is visually

connected with some and also connected with others beyond by air. Communication is prompted among the viewable rooms by the strong connection of visual element, and with the rooms out of sight by the senses other than visual sense. Subsequently, light and wind streaming into a room, as well as the act and the sign of the family there are transmitted to the adjacent rooms, and are extended beyond.

Supplementary Architecture

The rooms expanded in the site function as a house without being isolated functionally and spatially. The important thing is that the rooms are connected. The "connection" is formed by the persons and nature, and is not limited within the structure and the diagram of architecture.

I think that the architecture is something that acts as a supplement of the "connected air ".

1. Master Bedroom   5. Kitchen
2. Bedroom   6. Living
3. Study Room   7. Entrance
4. Bathroom   8. WC

N

PLAN 1:100

Ground floor plan

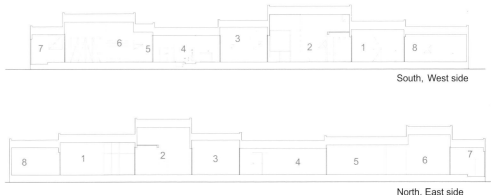

South, West side

North, East side

Section

盆唐Unjung-dong住宅
Bundang Unjung-dong House

Location Gyeonggi-do, Korea
Site area 269㎡
Bldg. area 131㎡
Total floor area 298㎡
Architecture design Kun-Jung
Architects & Engineers Associates /
Park Eun-kyu, Yoo Jae-sung
Interior design LIVINGHOUSE / Jang
Ji-ho
Photographer Lee Joong-hoon

"Bundang Unjung-dong House" is located in the residential complex within Pangyo Residential Site Development Zone and has pleasant surrounding environment. Buildings are built to give a feeling of soil and structured in a height with which people can completely experience five senses. Being fed up with the height of existing houses is inversely solved by lowness. From old times, the surrounding area of an old tree trunk base was used as a rest area for trees, stones and people. The shining of the sun much more emphasizes the role of a tree trunk base as a rest area. It was designed as a nature-friendly space which is in harmony with nature based on a design concept of "tree trunk base" and sunlight was secured by enlarging the size of the window in order to utilize natural light to the maximum.

Front elevation

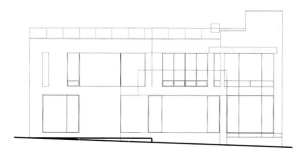

Rear elevation

Left elevation

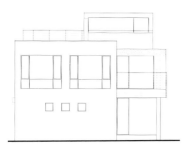

Right elevation

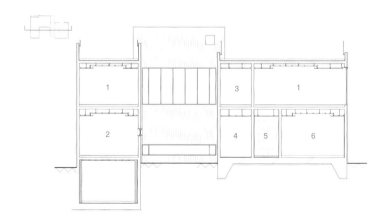

Longitudinal section

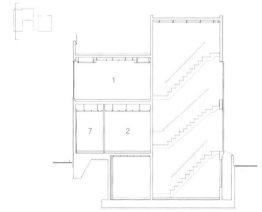

Cross section

1. Bed room
2. Dining room
3. Corridor
4. Entrance
5. Bath room
6. Master bed room
7. Utility room
8. Entrance
9. Terrace
10. Living room
11. Kitchen
12. Dress room
13. Family room
14. Porch

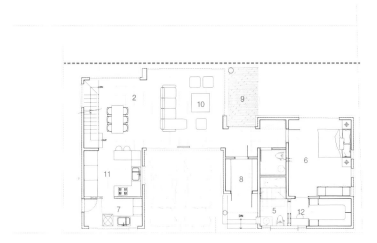

First floor plan

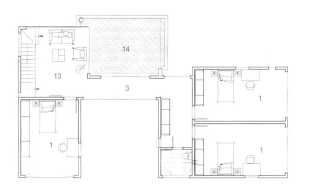

Second floor plan

Lasrado住宅 **Lasrado Residence**

Location Bangalore, India
Architecture design Cadence
Design participation Smaran Mallesh,
Narendra Pirgal, Vikram Rajashekar
Photographer Claire Arni

The house was to be designed on a tight plot measuring 30'x
40'. The intention was to create maximum open space on the
inside while the building has a distinct identity on the outside.
The bedrooms i.e. the private spaces were pulled out towards
the periphery to create an open space occupied by the common
areas i.e. the public spaces of the house. The open space in the
center was further articulated in levels as a series of platforms
that over look each other to ensure maximum visual and spatial
interconnectivity. The façade was seen as an independent skin/
mask that wraps the overlapping platforms on the inside. The
façade was also conceived as a series of concentric layers,
which created a strong visual identity in the neighborhood.

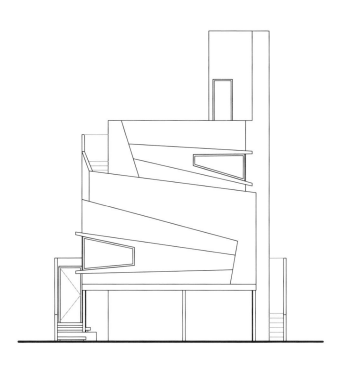

West elevation

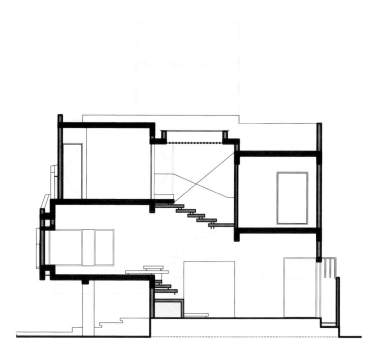

Longitudinal section

1. Parking
2. Foyer
3. Living
4. Dining
5. Kitchen
6. Bedroom 1
7. Wardrobe
8. Toilet
9. Family Room
10. Master Bedroom
11. Balcony
12. Toilet

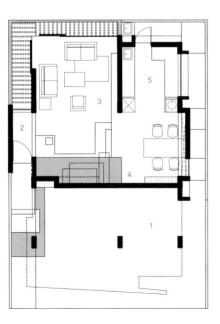

Ground floor plan

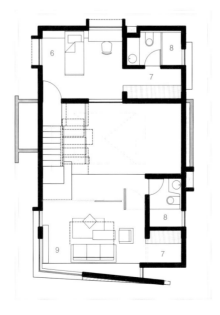

First floor plan

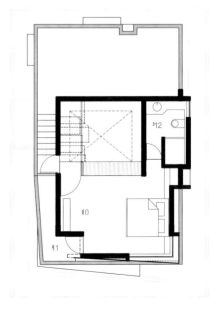

Second floor plan

达维史阿伯德别墅 Villa in Darvishabad

Location Darvishabad, Iran
Site area 400㎡
Bldg. area 120㎡
Total floor area 240㎡
Architecture design Rai Studio / Pouya
Khazaeli Parsa
Photographer Mohsen Jazayeri,
Mehrdad Emrani

The house is located in a small village in north part of Iran, near the Caspian Sea. A humid green land which is in a high contrast with the weather of Tehran and because of that reason it's a location for people who live in Tehran to spend their weekends. This project is a spatial dialog between traditional Persian architecture and aspects of modern architecture. It creates an architectural space through the incorporation of modern architecture tenets with traditional Persian architecture. To explain it in this house we have free space on the ground floor as the free space exists in modern architecture. On the other hand we have a kind of space surrounded by walls on top. This space opens toward the center, opens in a vertical direction toward the sky and the ground, a quality that you can find almost in all types of Persian architecture. The interesting part is the space in the middle, in fact as this modern space on the ground floor and this Persian space in the second floor start a dialog, a new space born in the middle. This space has got both qualities at the same time but is completely different and has got its own personality.

■ Design process

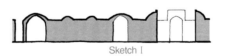

Sketch I

Sketch II

■ Structure

B   B

B   A

Sketch III

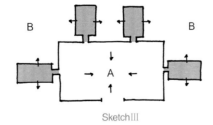

A

B

A

B   B

Sketch IV

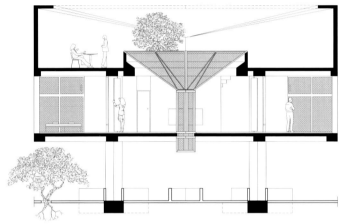

Longitudinal section

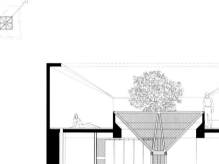

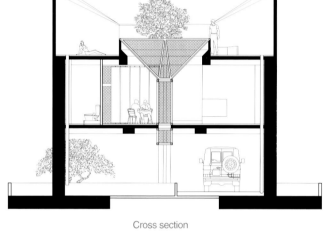

Cross section

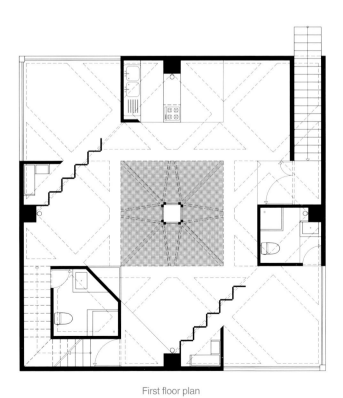

First floor plan

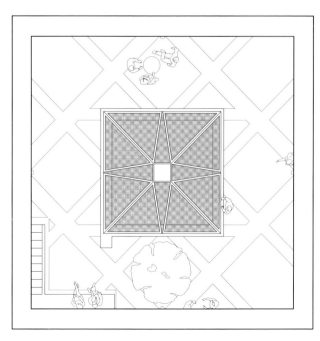

Second floor plan

韩国保宁住宅 **Boryeong House**

Location Chungcheongnam-do, Korea
Site area 985㎡
Bldg. area 182㎡
Total floor area 253㎡
Architecture design Well House
Architecture

The project site was an agricultural land located in Boryeong, Chungcheongnam-do. The original area was 198㎡ which was increased to 231㎡ after the design pre-review.

The front side of the land axis faces south, where An-san Mountain is located nearby. There is also Oseo-san Mountain with beautiful natural topography on the rear side. The project site gives comfortable feeling in terms of feng-shui (geomancy). Ansan and Josan in southwest are in good harmony with surrounding hills giving great view of sunset.

Shared spaces such as the living room, dining room and kitchen are located on the ground floor. Bedroom for the elderly mother of the owner is also located on the ground floor. Children's bedroom and family room are on the upper floor.

The owner had asked to let the design axis face exact south; however, it faces southwest because of the burden of too close An-san Mountain, which is just 500m in straight line.

An observatory terrace is provided on the upper floor. A great view to southwest is available from the terrace and the family room. The rear side staircase is designed with curtain wall so that a view to Oseo-san Mountain would be available while moving on the stairway and the south-north entrance axis will have connectivity through the staircase.

1. Entrance
2. Corridor
3. Living room
4. Kitchen
5. Dining room
6. Room
7. Bathroom
8. Dressing room
9. Terrace

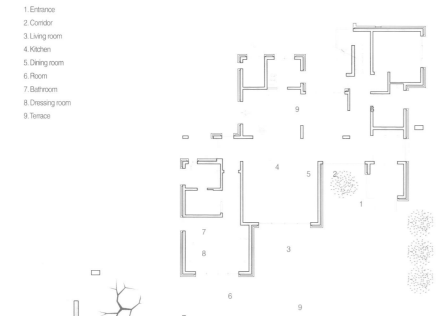

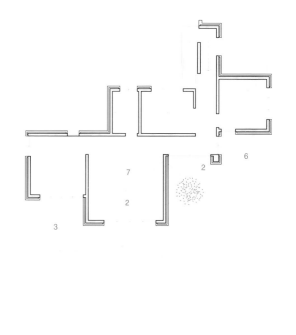

First floor plan   Second floor plan

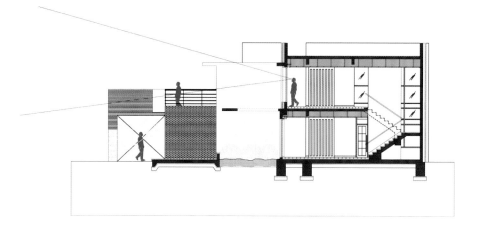

Section

JINO住宅 JINO HAUS

Location Gyeonggi-do, Korea
Site area 331㎡
Bldg. area 161㎡
Architecture design
PHILLIP ARCHITECTS Inc. /
Lee Ki-ok
Design participation
Park Sang-dae(Project manager),
Lee Seo-ho
Photographer Yun Jun-Hwan

The project site is a land for store/residence in Pangyo New Town. Its west boundary faces an independent residence block and the easy boundary faces 20m pedestrian-exclusive road. We decreased the rental space from two tenant households to one household and let the owner household use one and half floor. Before, 40% of the building area had been used as neighborhood living facilities; however, we moved some of the facilities to the underground.

The new program now has two floors of neighborhood living facilities, one tenant household and double-story owner household. The movement-line entering the upper residence made a long building mass from east to west. By this, the south naturally secured an open space which consists of a parking lot,

an outdoor deck and pedestrian path. The open-space axis does the communication role between the independent residence block and the pedestrian-exclusive road so that pedestrians would be led to the neighborhood living facilities. The 1st floor neighborhood living facilities support the upper mass only by cylindrical columns and they add openness by linking with the outdoor deck in the south. The underground floor is directly accessed through a sunken space. The outdoor stairway leads to the gate of the tenant household and the gate of owner household. The stairway is linked to the west road, of which the level is higher than the pedestrian-exclusive road in the east by 1.2m. The vertical movement-line naturally leads people into the building. The new architectural space created by the change

in cross section lets us experience a new architectural space. The stairway toward the house, the oblique line of cantilever and the curves on the rooftop engage with each other to form a solid exposed concrete mass with narrow and long shape. Three small disparate masses and the mass floating in the air are interpenetrated.

1. Retail
2. Leased residential unit
3. Residential unit for client
4. Sunken

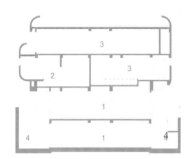

Section I

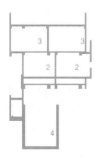

Section II

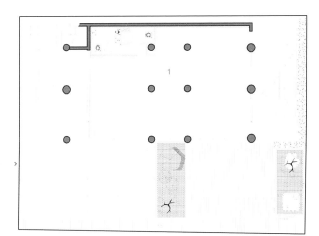

First floor plan

1. Retail
2. Leased residential unit
3. Residential unit for client

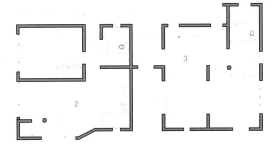

Second floor plan

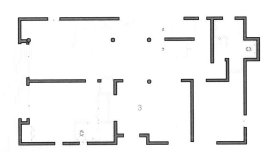

Third floor plan

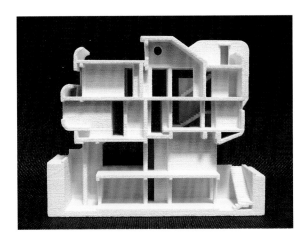

Section model

青浦区涵璧湾花园别墅 The Bay, Qingpu District

Location Qingpu District, Shanghai, China
Built area 19,495.9 sqm
Interior design Li Weimin
Architecture design Atelier Feichang Jianzhu
Principal architect Yung Ho CHANG
Project architect Liu Lubin
Design participation Wang Siuming, Liu Yang, Shi Chao, Qiu Yukui
Construction drawing China Shanghai Architectural Design & Research Institute CO., LTD
Photographer She He, Nic Lehoux

The Bay Garden is located at Qingpu District, Shanghai. The site was once used as a fishpond of nearly 43 hectares, with an ideal ecological environment. In the breeding season, a large number of water birds perched there. Our parcel is situated at the Island B, with altogether 20 houses of five types. The building areas aboveground vary from 514 sq. meters to 1022 sq. meters.

In our design, we try to bring together the architecture and the context. The latter is both natural and cultural — the water is a key element of the natural and the architectural tradition of the south is the prominent feature in the cultural. Meanwhile, the contemporary lifestyle and construction condition determine that the architecture will not be a mere repetition of the tradition.

Thus, the design was developed with a set of keywords: disperse, courtyard, and garden. Disperse — to take apart the different functions of a villa and then reorganize them into small groups. This move makes one building more of a combination of several buildings. In this way, more rooms are ensured good ventilation and day light which fits the humid and rainy climate of the locality and also blends the space inside and scenery outside.

Courtyard — the regrouped villa embraces several enclosed and half-enclosed courtyards of different sizes, providing the inhabitants livable outdoor spaces.

Garden — the landscape from the road to water introduces the residents a leisurely lifestyle. It also echoes the experience in a traditional southern garden. So far, each villa is a house as well as a miniature garden. People come here for the enjoyment of everyday life as well as sceneries.

In form, the gable wall and sloped roof reflect the traditional building elements in the vernacular architecture of the south whereas the untraditional construction materials — grey stone, aluminum alloy doors, windows and roofing, a steel channel that frames the wall, etc. — are a new interpretation of the regional heritage.

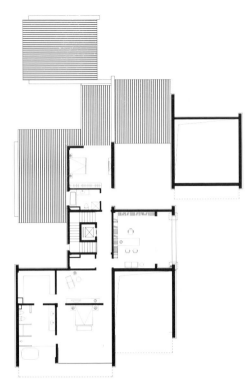

Second floor plan

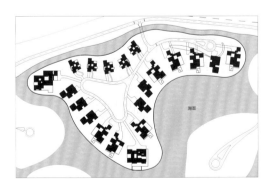

Master plan

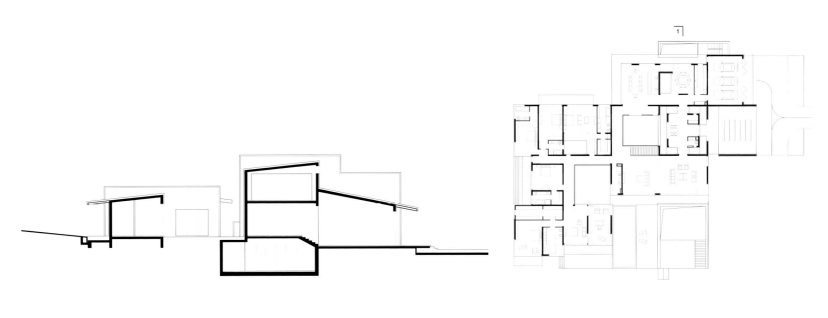

Section

First floor plan

Chohyangroo住宅 **Chohyangroo**

Location Gyeonggi-do, Korea
Site area 611㎡
Bldg. area 122㎡
Total floor area 251㎡
Architecture design Koossino
/ Koo Seung-min
Photographer Lee Joong-hoon

The place that faces Mountain Bara. Between the maintained boundary and the landscape, the only human interference that could have considered was the blank weeded space between the 7m revetment and the 10m revetment. The center of the residence area was leveled off along the contour of the accidented topography, and the buildings were narrowly aligned around the garden. The long, 7m revetment became the complete background of the outer side, and became a passway of the wind and the light making an adequate space.

To overcome the low building-to-land ratio, inner court was boldly excluded from the capacity. As the chain that links the interior to the open space, the stair took an important role. By tweaking the simple axis of the stair opened toward the outer space, the architect made some flexible spaces, which in turn contributed to the saturation of the abundance of the space by linking to the courtyard.

The courtyard that faces through the basement and the upper floor was planned to fully interact with nature. Centering the courtyard that preceded any other structure, the living space was bilaterally distributed.

The interior passage that meets the courtyard is narrow. By installing a simple stairway and allowing the light through the wall and ceiling that forms acute angle, the architect minimized the narrowness of the space. To convert the space's role from the mere passage to the courtyard, we temporarily opened the space both vertically and horizontally. Removing

the cumbersome setup, we opened the complete garden at the space that links between nature and the living space. We tried to minimize the feeling of material, wishing the space that can be incorporated into nature as the territory of exploration rather than that of the occupation.

As the record of the space that embraces the fragrance of the grass made from the goodness of nature, we hope Chohyangroo can revive as the place where mental meets nature.

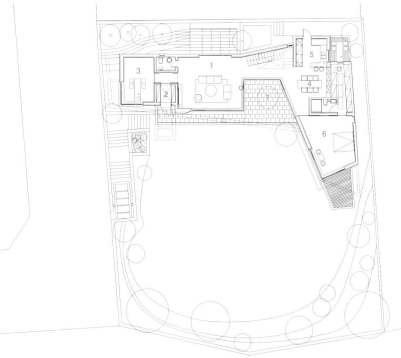

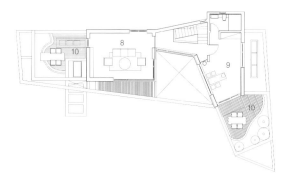

1. Living room   6. Master room
2. Entrance hall   7. Courtyard
3. Tea room   8. Library
4. Dining room   9. Guest room
5. Kitchen   10. Garden court

First floor plan   Second floor plan

日本水户住宅 House in Mito

Location Ibaragi, Japan
Site area 183㎡
Bldg. area 58㎡
Architecture design Takeshi Hosaka
Architects
Photographer MASAO NISHIKAWA

This architecture is composed of stacked levels of inside and outside.

At northeast, there exists native forest, and 15-storey-high apartment at southwest. The Mito Residence is located right next to the conventional and stiff mass dwellings, and proposes a living of inside and outside.

The levels are composed and stacked in a way that creates forest-facing outside by anchoring northeast corner and setting each level back from opposite corner. Consequence is that the building gets smaller as it rises.

This eased the presence of huge apartment and the eyes, whether being outside or inside.

This house spears somewhat closed off from its surrounding context at first glance. Entering the ground level that is solely inside. As one moves up to the smaller first level, outside appears. Similarly loft level has both inside and outside, and another level above is rooftop where inside no longer exists and the space becomes entirely outside. Towards the sky, built volume gets smaller but openness increases at the same time. A journey through the ground level entry to the rooftop, an outside proportion increases and eventually becomes the sky. Having a sense of being in the house, inside and outside coexist and unfold as journey continues.

The merged space of inside and outside is an unified architecture of their characters and qualities. Reading books, brushing teeth while strolling around, taking afternoon nap,

painting, watering plants at various spots, reading newspapers, playing with toys, or talking on a phone, spaces are occupied and used, may this be inside or outside, as one feels like or by absolute coincidence.

This architecture differs from those contemporary architectures that identify the inside and outside and allocate functions as to organize how man is to dwell. Instead, this architecture aimed to accommodate various active living ways that spread across inside and outside.

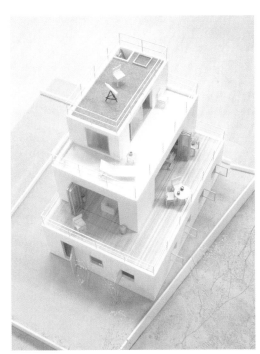

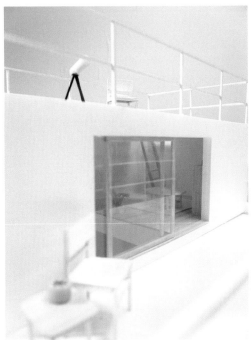

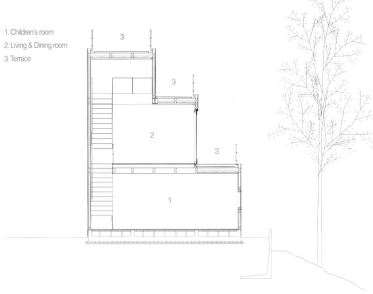

1. Children's room
2. Living & Dining room
3. Terrace

Section

1. Entrance
2. Lobby
3. Toilet
4. Bath room
5. Children's room
6. Bed room
7. Cloth room
8. Living & Dining room
9. Kitchen
10. Terrace

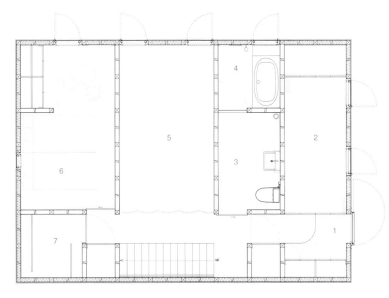

First floor plan

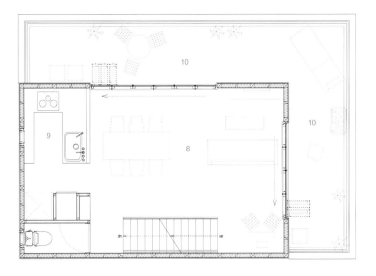

Second floor plan

HyoJae住宅 **HyoJae**

Location Gyeonggi-do, Korea
Site area 2,045㎡
Bldg. area 286㎡
Total floor area 264㎡
Architecture design STUDIO
KOOSSINO / Koo Seung-min

Architecture is developed and refined based on the layers of topography, which is called nature. The high-low of topography and the big-small of area divide the mass-space into small pieces. Then the pieces are stacked in multiple layers occupying the space. In other words, the space consists of small pieces made from the relationship of density and occupying that topography. The land wraps the flow of invisible topography. The generation cycle of it is the sediment of the temporal space. Hyojae is a space created in variety by dividing the stacked layers of topography by each layer.

Hyojae is on the low ridge at the entrance of New Seoul Golf Club located on the way from Seongnam to Gwangju. Hyojae is the pen name of the Owner meaning industriousness. However,

the image from the first impression of the site was the shape of [the original staying being]. The outlook of Hyojae with the roof in geometric shape is against the mountains; however, the arrangement of the building is laid out with balance smoothly. The shared living room is located in the location with most sunlight. Shared rooms for the lady of the house and guests are located to the left and right of the living room. The horizontal volume of the building settled down on the land and the lifted vertical space of the living room counter-pose with each other to make the space arrangement more conspicuous.

The house has a building height of two story house by outlook; however, it is a complete single story house by establishing its mass combination relation. It is also a space creation on the

upper level relation. The spaces of the main building and the detached building, which are linked to the axis of the flow, are divided by the skylight. A corridor with rich sunlight is located in parallel with the straight canopy next to the upward path. This bright corridor does the role of a courtyard. The design established a space loop of opening and closing so that overall outlook will agree with the order of nature. The horizontal band of glass layer has a neutral order and harmonizes the whole. The cold Marblelite counter-poses with nature creates richer texture and the silhouette changes in accordance with the flow of light.

I just hope the house will be a solid object like an existing stepping stone in nature rather than an architectural object.

Sketch

1. Entrance
2. Living room
3. Dining room
4. Kitchen
5. Multi room
6. Room
7. Bathroom

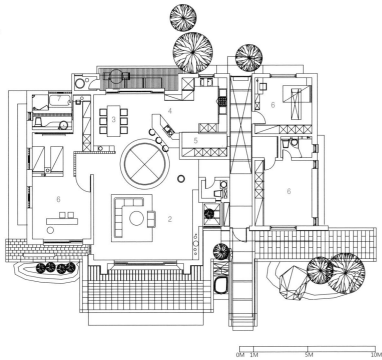

0M 1M 5M 10M

Base floor plan

0M 1M 5M 10M

Front elevation

0M 1M 5M 10M

Section

Z住宅 Z-House

Location Gyeonggi-do, Korea
Site area 805㎡
Bldg. area 190㎡
Total floor area 288㎡
Architecture design Hohyun Park +
Hyunjoo Kim / Park Ho-hyun,
Kim Hyun-joo
Photographer Seok Jung-min

The site is located at Gwangju si, Gyeonggi-Do, where is not so far from Bundang new town. Surrounded by small houses, it is at the top of a hill with deep slope.

First thing to consider is to keep woods on the west side of the site. Available area of the site is limited by the woods. Facing a view of mountains on the south side, the building is located. In composition of space, relationship among programs and circulation played an important role. Ground level, which is divided by kitchen/dining and living area and upper level, which is divided by children's rooms and master zone is crossing at division area. By this manner, space is gradually ascending from entry to master zone. Next thing to consider is to make dynamic space by changing size of space and by leading sight. High ceiling at the entrance corridor is more emphasized by sunlight through skylight and a wall, which is faced at the end of corridor, inducts eye to dining area. At this point, the ceiling height is suddenly changed by overhead ramp. High ceiling living area appears through the low-rise 3 step staircase, which is crossing outside water pond and inside plant area. From living area, space bifurcates to library at lower level and to upper level. Master zone, where is a climax in space scenario, is connected by a corridor and a ramp from children's room at upper level. Master bedroom is entered through open bathroom and powder room and reconnected to entrance corridor.

Shape of the building is planned to follow the space scenario and roof, which is covered by black zinc is wrapping around upper level mass and form a homogeneous and muscular shape. Contrarily, lower level masses, which are covered by basalt, stand rigidly. Water pond and inner plant space are inserted at the crossing point of lower level mass and upper level mass.

(Written by Hohyun Park)

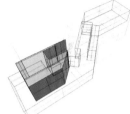

Mass

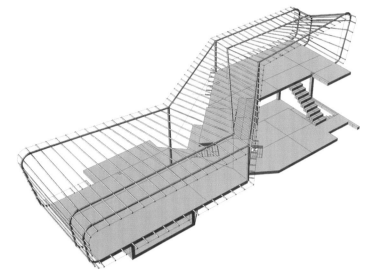

Structure

1. Deck
2. Living area
3. Corridor
4. Children's room
5. Library
6. Laundry area
7. Terrace
8. Master bathroom
9. Kitchen

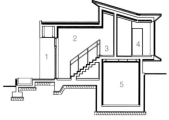

Cross section Ⅰ

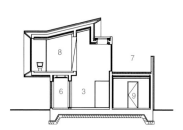

Cross section Ⅱ

1. Walk-in closet
2. Master bedroom
3. Master bathroom
4. Living area
5. Terrace
6. Ramp
7. Children's room
8. Bathroom

1. Entry porch
2. Entry
3. Corridor
4. Laundry area
5. Kitchen
6. Dining room
7. Bathroom
8. Pond
9. Plant area
10. Living area
11. Deck

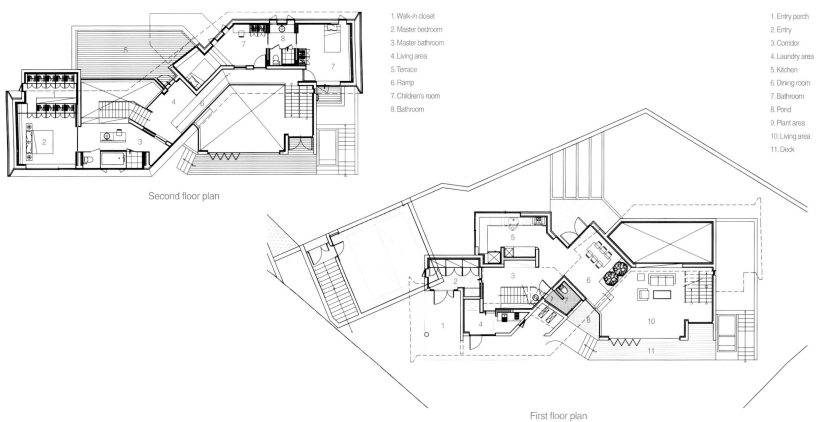

Second floor plan

First floor plan

OGAKI住宅 OGAKI House

Location Gifu, Japan
Site area 130.64㎡

Built area 74.54㎡

Total floor area 106.42㎡
Architecture design Katsutoshi
Sasaki + Associates
Structure company Masaki
structural laboratory
Construction company Tamada
construction Ltd
Photographer Toshiyuki Yano

This residential house is introduced from the ambient surrounding and the conditions of site planning. In winter seasons, the strong west wind (the fall wind of the Ibuki) blows in this area so it is the plan of suppressing the load of the building by extending its roof up to the close to the ground soil and fending off cold winds at the roof.

In summer seasons, it is the structure that discharges the accumulated heat inside to the outside through the void of the inner court and the central void installed various parts. The openings face only to the courtyards in the east and west of the site and there is none in the north and south.

This way heat penetration in summer and heat loss in winter via the openings can be controlled. Living rooms and patios are arranged in turn inside the triangle volume, which makes an explicit proposal about closeness between life and nature. However, the most important feature in the house design should be how communication among family members can be assured.

For that matter, the design has many elements to encourage communication, for example, a kitchen facing the entrance (to provide an opportunity for a mother to meet her children when they come home), a children's room spreading throughout the second floor (four daughters share it), and an open ceiling space (a louver type floor) which the voices of family members can get through.

Considering the site context, external elements, heat environment, family communication, structure plan and cost, I decided to select a triangle design.

1. Bed 01
2. Lounge
3. Void
4. Bed 02
5. Garden

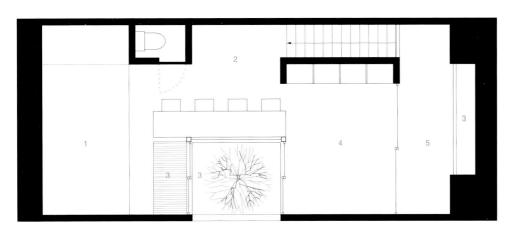

Second floor plan

1. Storage
2. Pool
3. Bath
4. br/a & Lounge
5. Kitchen & Dining
6. br/b
7. Living
8. mbr
9. Garden

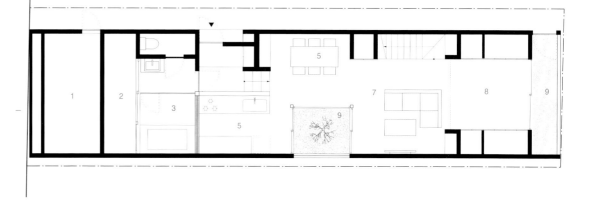

Site & First floor plan

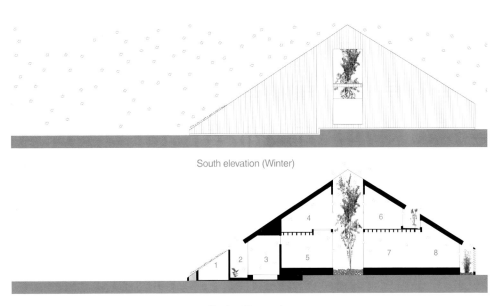

South elevation (Winter)

4
6
1 2 3 5 7 8 9

Section (Summer)

1. Storage
2. Pool
3. Bath
4. br/a & Lounge
5. Kitchen & Dining
6. br/b
7. Living
8. mbr
9. Garden

交流之家 Communication

Location Seongnam-si, Korea
Site area 814㎡
Bldg. area 162㎡
Total floor area 596㎡
Architecture design A-one Architects
Consulting Engineers Ltd. /
Lee Dong-hyun +
S. J. Development / Lee Keon-ho
Photographer Kim Myoung-sik

Geumgokdong, in which the house is situated, is a village where residences toward calmness and serenity are beginning to emerge, out of a city. In this village of fragrance of grass and birds' singing, the heat of a large city and front road seem faint.

3, 3, 3

The main concept of this house is 3-3-3: it's the third house of the owner, while 3-floor building is demanded, which 3 generations could stay together. As is the third house, its material selection was proceeded scrupulously for a better residence than the former. Basement floor is allocated for the owner, ground floor for a family, second floor for their children and grandchildren. Considering for poor mobility of the owner

in a building of 1 floor below & 2 floor above the ground, home-use elevator is installed, while slope and height of stair were minimized for maximum convenience of the owner.

Serenity of winter

Blazing summer had passed... Autumn, and the next, winter... As an architect, I wanted to embody serenity in this building. Soon a grandson would come to the house. The owner, who was a strict grandfather as the president of an enterprise, might hope to roast sweet potatoes on a fire in the white winter, feed it to his lovely grandson, and watch the cute baby, contentedly. As his early life was fierce, this longed-for peace would be much dearer for him.

Communication with nature

This house breathes with nature. It embraces topography of nature as it is. The flow of mountain is introduced directly into a garden. The sunken garden draws a sky and nature into the interior, underground, and the sunshine penetrates rooms as is on the ground floor. South-facing windows on the 1st & 2nd floor embrace scenery and sky of the hill, impressively, making a house look like floating in the sky.

The time of the old age, which the young come to imagine at least once; everyone would like it to be a time of smile. Nature, family, love... I wish the owner would feel these peacefully in this house.

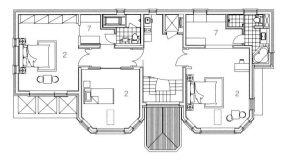

1. Living room
2. Bedroom
3. Kitchen & Dining room
4. Drawing room
5. Library
6. Multipurpose room
7. Dress room
8. Terrace
9. Parking lot

Second floor plan

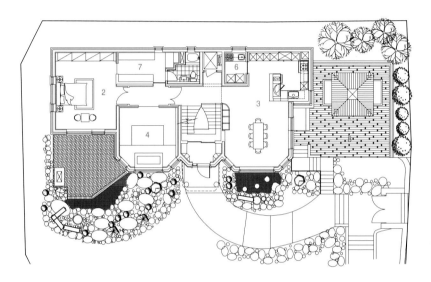

First floor plan

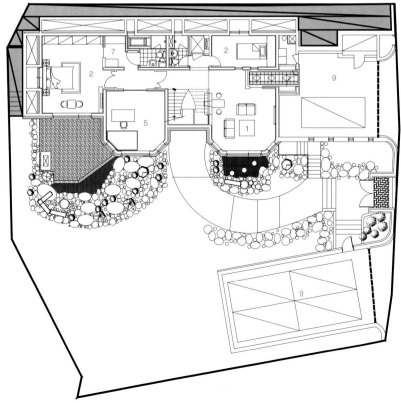

First basement floor plan

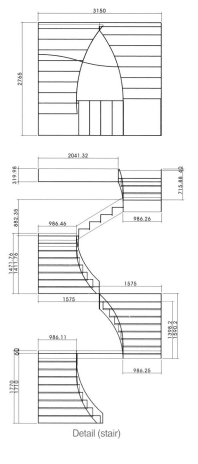

Detail (stair)

Front elevation

East elevation

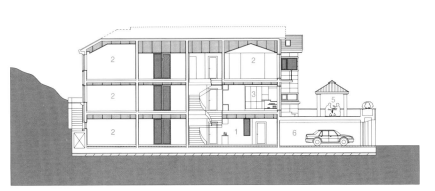

Cross section

1. Living room
2. Bedroom
3. Kitchen & Dining room
4. Drawing room
5. Terrace
6. Parking lot

Herzelia Pituah住宅4 Herzelia Pituah House 4

Location Herzelia Pituah, Israel
Plot size 1,000 m²
Built area 600 m²
Architecture design Pitsou Kedem in cooperation with Tanju Qzelgin
Design participation Pitsou Kedem, Noa Groman
Photographer Amit Geron

An open but enclosed home

A modern and minimalistic building that completely eliminates the accepted barrier between the inside and the outside and hides, within its restrained spaces, a dramatic atmosphere and wide open spaces.

The house was built as a vacation home for a family living abroad. It is situated directly on the coast in the center of the country. The building's architectural design is based on three central masses that surround a large internal courtyard with a swimming pool at its center. The masses comprise the border and the barrier between the street, the neighbors and the home's interior and between the internal courtyard and those same spaces. The central theme was to create dynamic walls that allow, on the one hand, the elimination of the boundary between the central courtyard and the internal spaces, and on the other hand, the creation of a changing and dynamic facade that allows for the total closure of the façade or different levels of exposure or concealment.

The central mass is bounded on the side facing the pool, by a seven-meter-long aluminum display with no supporting columns. Half of the display is divided into six large continuous units that are embedded in wall niche and are, in effect, concealed units. The encompassing mass becomes a floating one and the boundary between the kitchen and the long bar that extends along its entire length and between the pool and the surrounding deck is cancelled out in its entirety and so creates an open home with no boundaries. A space where the border between the inside and the outside becomes blurred and, in effect, creates a new space that is a cross between an internal and external space.

The side, the sea to the west, allows for a richer relationship between the exposed and the concealed. A partition created from six equal metal sections and adjustable wooden slats allows for the total closure of the façade or different levels of blurring between the inside and the outside. The dynamism of the large accordion doors and the adjustable wooden slats

allows natural light that enters through them, to play a central role in the home's internal spaces and for the creation of geometrical displays of light against the vertical and horizontal areas and so enrich the material restraint in which the home's spaces are fashioned.

The architectural idea comes to the forefront in a concrete fashion when all the building's facades are opened and disappear. The structure, which looks like a solid and sealed structure, suddenly appears to be floating and produces a feeling of lightness and visibility. The house appears as if it were built from floating surfaces, with long moving pictures that divide and move between them.

The central motives that usually accompany the architect: reliance on fewer shapes, a reduction in materials and restrained colors are manifested in this project. But, what makes this project special is that, in order to strengthen the dynamics and the tension between the faultless and restrained masses and between the feeling of lightness and openness when all the walls are opened, two central materials are used that are the opposites of each other: solid, strong and restrained materials such as grey limestone and black basalt as opposed to clear and opaque glass and other reflective and reflecting materials such as water in a black reflective pool. The dialogue and the relationship between the materials and the meditative qualities, create, within the restrained building's spaces, a unique and surprisingly dramatic atmosphere.

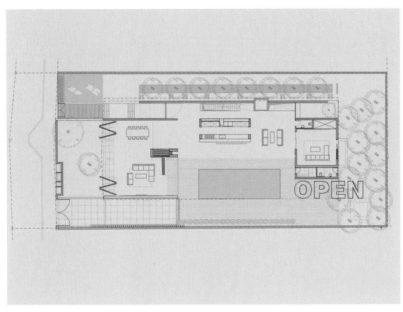

Plan 1

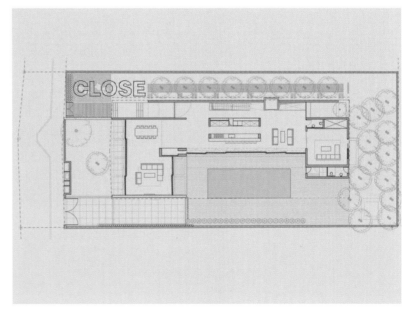

Plan 2

Model plan

Rao住宅 The Rao Residence

Location Hyderabad, India
Site area 795 m^2
Built area 1,200 m^2
Architecture design DDIR
Architecture Studio, Pvt. Ltd,
Bangalore, India
Design participation Dominic Dube
(principal architect), Inge Rieck
Photographer Bharath Ramamrutham
Consultants Furniture, Furnishings
– MDF Milan, Paola Lenti Milan /
Kitchens – Poggenphol / Lighting –
ALC Chennai / AC – Toshiba / Security
– Home Automation Hyderabad / Pool –
Olympia Pool / Elevator – Domus Lift
Italy / Windows – Veka Windows /
Landscape – Padma Rao

Vision

The clients' requirements were nothing but sublime and serene, a beautiful inherent quality to the Ramakrishna philosophy and to the Vastu principles. We are searching through for a new vibration, a new feeling of life combining those two elements and we plan to achieve with the correct balance of the overall planning around and within the residence. A simple creative spirit must be laid over the entire residence in order to offer a fresh environment for the users. The buildings and connecting elements - the staircase, the lift, gardens and walls - must work together as a poetic whole, borrowing and returning views and vibrations from each other in a constant symbiotic play of form, space and light.

Simple and appropriate decisions need to be taken to transcend the philosophy and offer something distinctive and both in its visual presence and the lifestyle within.

View / Connections

The ensemble and landscape should be integrated in a way that is poetic and sensitive enough to match the stunning requirements. To achieve this, the visible and perceived connections between the various elements of the house are most important. The architecture must organically melt with the surrounding elements and nature and create a rich and beautiful life.

Mass / Architectural Spaces

The mass, a perfect cube created on mathematical proportion must be one of visual continuity between all functions of the house, inside functions and outside loggias and terraces and is accomplished with a smooth flow of spaces all playing with shadows and light. Spaces, in and out, landscape and built space, are there to foster the feeling of continuity and peace.

Passage

Walking along the house and through is like being energized and animated by the light, shadows and breezes whispering through the spaces, the water and the feeling of the entire mass and landscape is a continuous link dedicated to meditation.

Gadrens

The built landscape and the gardens stand alone and still live up to their potential. Without an understanding of the architecture, the overall landscape design can lose its nuance and power. The two factors need to work together and be integrated with the intelligence and understanding of the contextual nature and specificity of the concept. The plasticity is then born in great harmony.

We are proposing a series of private gardens, influenced by the styles of the residence and melt with the inherent feeling created by the philosophy of Ramakrishna. Infinite variations for these little gardens will create many intimate pockets and niches.

Structure and soul

To contain the entire project a simple exposed steel structure has been designed on a mathematical grid to assure the rigor and playfulness of the concept. The columns will be of a cross shape and will be as sculptures all over the residence wherever they will be. Rather than giving the feeling of hiding something, as most architecture do, the structure gives the feeling of continuity in permanent dialogue with the users. This living structure transports you into the world of delicate lights and divine shadows; it makes you feel as if you are the philosophy itself.

Lifestyle

The lifestyle this residence makes possible is one of simple elegance in a personal retreat, in touch with nature, light and oneself.

We've created beautiful atmosphere for contemporary life.

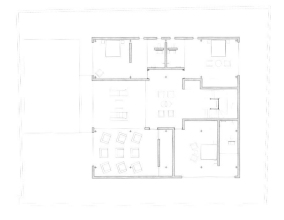

First floor plan

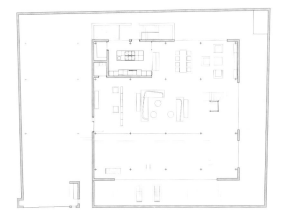

Ground floor plan

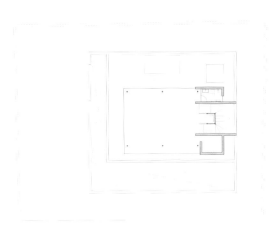

Terrace floor plan

G形住宅 G-house

Location Gyeonggi-do, Korea
Site area 660㎡
Bldg. area 178㎡
Total floor area 197㎡
Architecture design Hwang Jun
Architects & Associates / Hwang Jun
Photographer Park Young-chae

Building a new country house

The building is located in a country house district in Gapyeong. There are numerous ordinary houses made of materials like red brick, siding and wood. Despite their classical appearances, the interiors are not so much different than those of existing apartments: a large living room with a fireplace standing all by itself.

Since people have different tastes and the trends are always on the move, there is no such thing as an ideal way to build a house. Yet, I could not help but think that many country houses in Korea failed to interest their inhabitants. And then I was given a chance to design one in a mountain in Gapyeong.

The site is located on the slope in the south-eastern part of

the mountain. In front of the building runs a stream. Because the entire site is slanted, the biggest challenge is to ensure a large front yard without reclaiming the stream. However, the limitation of the site can also turn out to be a good opportunity for a designer to experiment a new plan. Making the best of the given site conditions, I try to build a country house that is unlike the others.

The spatial composition includes three bedrooms for the owner couple and two children, a living room and a dining room/kitchen. The exterior of the house was simplified at a maximum extent. The living room is shaped as a simple box with a high ceiling. Considering the harmony with surroundings, glass and copper plates were used to finish the exterior wall, and the birch

plywood was applied to the interior. Not only the walls but also furniture, sink and doors were all unified by a single material: birch.

The entrance in the back of the building opens to the corridor that leads to the living room where one encounters the outdoor landscape lying beyond the windows. To enhance the dramatic effect, we designed the upper part of the living room as a glass box, and removed concrete beams usually found in the front part of the space.

The glass flooring of the terrace offers the view of the stream, the rooftop veranda opens up to the sky and the inner glass wall allows the natural light to come into the interior... the inhabitants of this house should be able to find trivial but amusing things in their life.

This "extraordinary" country house could never have been built without the agreement and trust of the owners. Since they had already lived in one of those existing country houses, they must have been able to free themselves from fixed concept.

Now that I see the house realized, it seems to harmonize with surrounding environments much better than I thought.

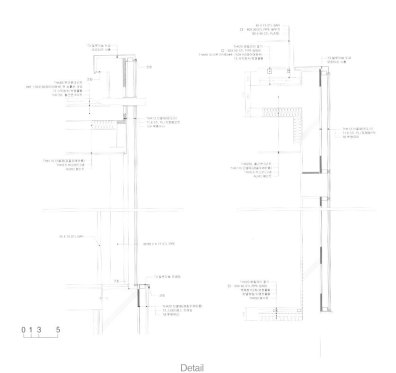

Detail

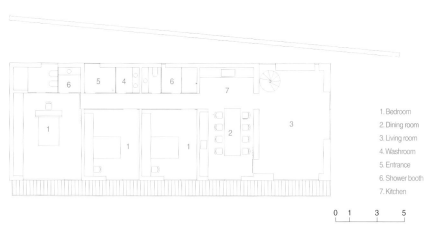

1. Bedroom
2. Dining room
3. Living room
4. Washroom
5. Entrance
6. Shower booth
7. Kitchen

0 1   3   5

First floor plan

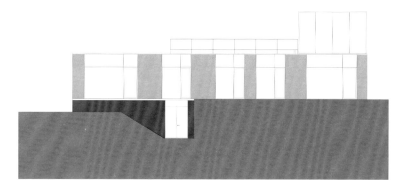

Front elevation

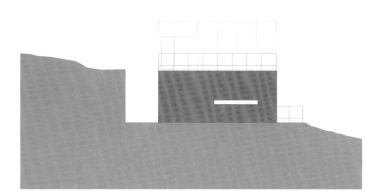

Left elevation

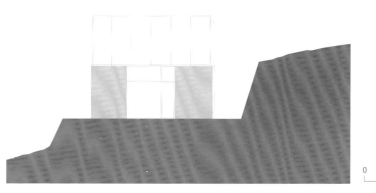

Right elevation

0   1   3   5

GWAN RYU HEON住宅 GWAN RYU HEON

Location Kyeongsangbuk-do,
Korea
Site area 793㎡
Bldg. area 117㎡
Total floor area 95㎡
Architecture design A.M Architects
/ Kim Tae-yun
Photographer A.M Architects

Gwan Ryu Heon is the house for contemplation of "the flow of metaphorical architecture" through a blank space between physical border and nature. The eyes pass an "empty frame" in an achromatic color, which is stood at an entrance. They pass another floating "filled volume" in a dramatic scale with picturesque patterns and enter inside. At that moment, they feel the diffused light of nature that comes down from the ceiling. The light streaming at a dynamic angle and the flow of a long path are permeated into a glass wall and organically connected with exterior decks which spread successively. Then, they gradually go toward the nature again through a transparent frame and look into the flow. A horizontal fake wall that goes with the flow of nature, together with a vertical wall that arouses

tensely, shows a dramatic contrast effect. They are permeated through empty volume and unveiled lines of nature opened successively. Proceeding in accordance with the flow of the fake wall, the movement of circulation is connected into nature after passing a hill on the border.

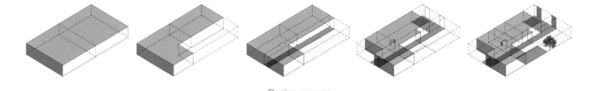

Design process

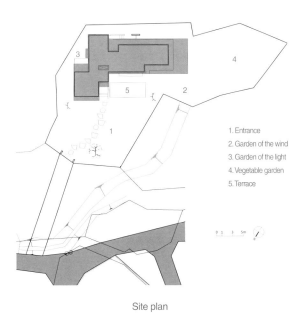

1. Entrance
2. Garden of the wind
3. Garden of the light
4. Vegetable garden
5. Terrace

Site plan

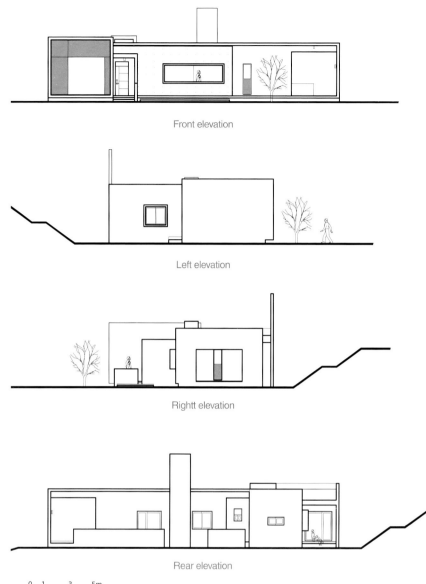

Front elevation

Left elevation

Rightt elevation

Rear elevation

0　1　　3　　5m

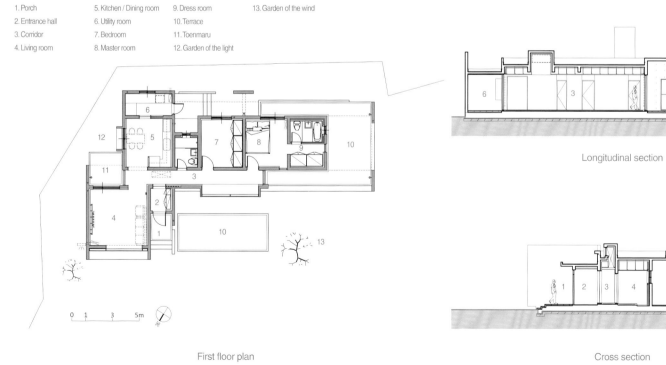

1. Porch
2. Entrance hall
3. Corridor
4. Living room
5. Kitchen / Dining room
6. Utility room
7. Bedroom
8. Master room
9. Dress room
10. Terrace
11. Toenmaru
12. Garden of the light
13. Garden of the wind

1. Porch
2. Entrance
3. Corridor
4. Bathroom
5. Service space
6. Toenmaru
7. Terrace

Longitudinal section

0　1　　3　　5m

First floor plan

Cross section

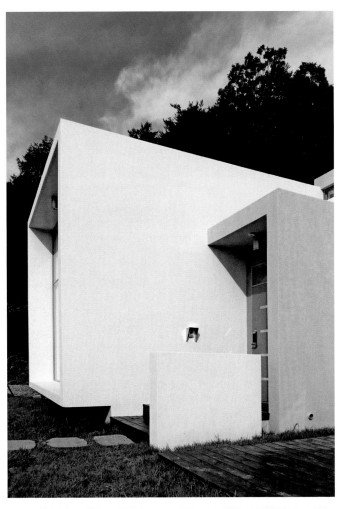

S形住宅 S-House

Location Seoul, Korea
Site area 654㎡
Bldg. area 191㎡
Total floor area 499㎡
Architecture design Hwang Jun
Architects & Associates / Hwang Jun
Photographer Lee Ki-hwan

This is a house for an owner couple and their three children. The building consists of an underground floor and two above-ground floors. The 2nd floor, where the entrance is located, is a public space composed of a living room, dining room, kitchen and a multipurpose room, and the 1st floor is for private spaces such as a master bedroom and children's rooms. The underground floor is occupied by a family room, miscellaneous room and a boiler room. The simplicity is maximized within the four corners of function. The main spaces are positioned in the south and the subsidiary ones in the north. Continuous open spaces were created in front and back of the building to bring in the daylight. The house does not show off from the start. The main concept was to look

somewhat plain and simple. All the elevations on the inside and outside were divided as necessary, without any lame attempts at innovative design. We tried to produce a different effect with usual finish materials. Rather than introducing unique and original ones shown in foreign countries, we used already popular and familiar materials such as bricks, blocks, glass and plywood, so that the building would appear more architecturally.

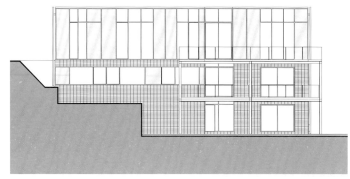

South elevation

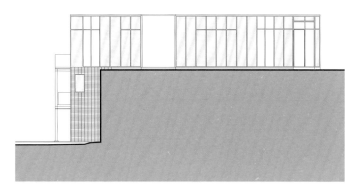

North elevation

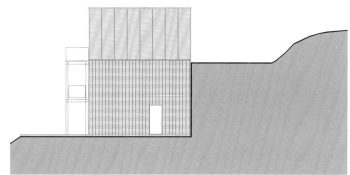

East elevation

1. Utility room
2. Kitchen
3. Dining room
4. Living room
5. Room
6. Family room
7. Bath room
8. Audio room

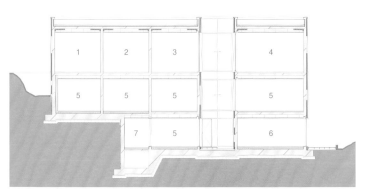

Section Ⅰ

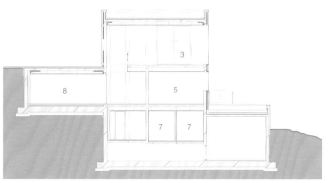

Section Ⅱ

1. Entrance
2. Living room
3. Dining room
4. Kitchen
5. Utility room
6. Waiting room
7. Balcony

Second floor plan

1. Room
2. Study room
3. Storage
4. Guest room
5. Balcony

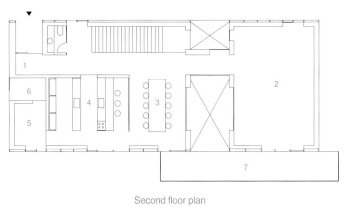

First floor plan

1. Living room
2. Room
3. Storage
4. Deck

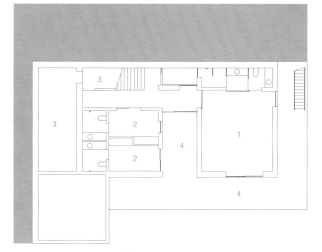

First basement floor plan

静默的住宅 House of Reticence

Location Shiga, Japan
Site area 164.29㎡
Built area 135.59㎡
Architecture design FORM/Kouichi
Kimura Architects
Photographs Takumi Ota Photography
www.phota.jp

This house is built on the triangle site with a width of 18m. The client has requested to make the best use of the characteristic site form to build a house with both privacy protection and a sense of openness in the house.

The building is composed of the echelon volume successive along the site form, and the high wall. The landscape-oriented façade, which is one of the external features and brought about by making good use of the site width, allows people's line of sight to be introduced in the horizontal direction.

The interior space design also takes advantage of the site width. On the first floor the entrance hall is located at the center. On its both ends are the spot gardens that are allocated in the spaces separated by the Japanese room on the irregular site form.

As the line of sight is designed to be as long as possible, the internal space is visually expanded so as to realize the space that gives an open feeling. On the second floor the living room and the balcony are laid out on both ends. In addition, the ceiling of the living room is designed to be higher than that of the other rooms. These designs intensify visual expansion.

The opening at the upper side of the living room, as well as the glass wall on the balcony where a bench is furnished, is one of the elements that produce a sense of openness.

By considering the site form to select the locations for the openings and control the line of sight, this house realizes the spaces that give a sense of openness but are closed off to the periphery.

1. Kid's Room
2. Hall
3. Bedroom
4. Japanese Style Room
5. Balcony
6. LDK
7. Storage

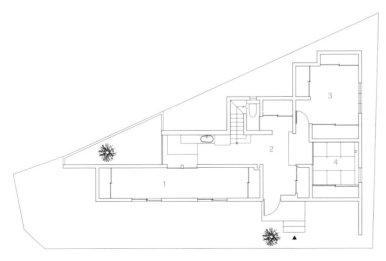

Ground floor plan

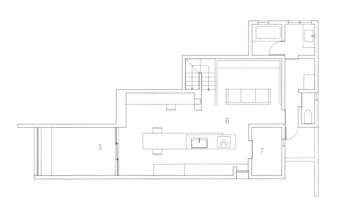

First floor plan

Sangye住宅 Sangye House

Location Jeju-do, Korea
Site area 817㎡
Bldg. area 144㎡
Total floor area 99㎡
Architecture design SIYUJAE Architects
& Planners / Ko Seong-cheon
Design participation Byun Young-suk,
Kim Young-beom

This project of a small-scale detached house begins from a simple box form which consists of a floating mass and a supporting submass. The floating mass was hollowed out in the center to bring nature into the interior. The exposed concrete and the contrasting Jeju stone with its timeless and strong character were selected as materials. The position of the layout is off the land axis for the sake of privacy, and the inside and the outside are linked by the balcony. The level difference with the exterior gives neutrality to the balcony to make it seem to belong neither to the inside nor to the outside. Such spatial hierarchy called for lifting the building from the ground, resulting in the image of a floating house. The completed building shows boxes floating in the air on different levels. The neutral space between the interior and the exterior allows the front yard to serve both as a private room and a public zone separated from the inside; the owner's family could have a chat in the living room, enjoying the view of the front yard and nature brought into the building.

1. Inner room
2. Garden
3. Living room
4. Room
5. Dress room
6. Bathrooms
7. Kitchen
8. Multi-purpose Room

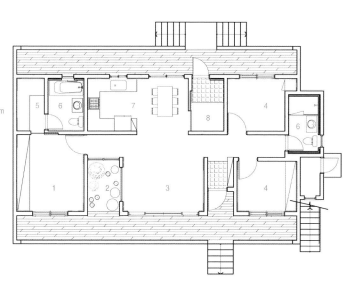

First floor plan

庇护住宅 Santuary House

Location South Jakarta, Indonesia
Site area 243㎡
Architecture design Atelier Cosmas Gozali
(PT. Arya Cipta Graha)
Designer Arch. Dipl. Ing. Cosmas D. Gozali, IAI (Principal Architect)
Setya Kurniawan, ST (Project Architect)
Photographer
Tectography (Fernando Gomulya)

The atmosphere is calm and comfortable in the neighborhood around this house which architects intentionally incorporated into the building and also a strategy to place the park on the first floor making this home has extensive views of the garden to the exterior of the building with no obstruction by surrounding buildings. The existence of the garden floor of this one apart is the point of interest in buildings, and also contributes to the green space in urban environments. Processing of traditional Balinese architectural concepts such as Wantilan and processed blank bale is an award to the wealth of contemporary Indonesian architecture that can present with more modernity. Wantilan, a modern form contained in the dining area is connected by a foyer into the living room area with

theorientation of openings to the garden area. Other Balinese architectural concept that is applied to the blank bale bedroom upstairs. Translation blank bale is processed in a modern form of usage rounded columns and walls of glass that seemed to directly support the ceiling. So that the room seemed transparent to the outside as if protected by a roof only.

The modern concept is also applied, starting from planning to layout compact space arranged to meet a lot of space on the limited land. Then the use of natural lighting during the day in every room, even for room service. Technological glass as a wall of light penetrating the fish pond to the hobby room on the ground floor is also a detail of lighting that is unique in this house. Aluminum frame material with the use of technology

that can reduce heat and noise from outside, as well as green glass that reduces heat from the outside without reducing the intensity of incoming light, and also planned openings that avoid the west in order to minimize the hot afternoon sun are a consideration key in creating an environmentally friendly residential and convenient for the residents.

In general, the facade of modern architecture houses is dominated by white color in order to give the impression of simple, modern, and abstract in buildings with limited land, besides the white color on the buildings as well as a blank canvas that will be filled by the color of the surrounding environment, shadow of buildings and trees, and light effects from the house itself. To adapt to its Balinese architectural concept, the material used in building facades is woodplank to give effect to the tropical timber. The combination of white and wood elements merge harmoniously in this building with a g reen and quiet environment so that the dwelling house is like a sanctuary in a very urban neighborhood. Broadly speaking, this house design is a modern interpretation of traditional Balinese architecture which is composed to suit contemporary needs and the times.

1. Foyer
2. Family Room
3. Dining Room
4. Kitchen
5. Powder Room
6. Play Room
7. Guest Bedroom
8. Garden
9. Main Bedroom
10. Dressroom
11. Child Room
12. Garage
13. Hobby Room
14. Storage Room
15. Service Area

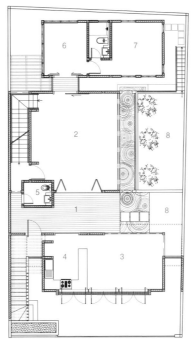

First floor plan

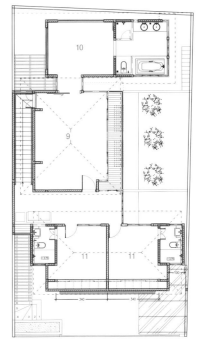

Second floor plan

Wakayama的住宅 House in Wakayama

Location Wakayama, Japan
Site area 175.19㎡
Architecture design Yoshio Oono
Architect & Associates
Photographer Kenji MASUNAGA
(Nacasa & Partners)

These are the plans of a married couple and their children's private house built in the quiet residential area in the Wakayama prefecture in Hashimoto city. One can feel closer to nature under the wide blue sky, which seems wider than usual as low-rise houses are clustered all around this private house. Having some living space that takes in as much nature as possible was one of the conditions of the housing plan. (The boundary between the inside and the outside of the house becomes vague.) In order to bring the outdoor elements and the indoor elements closer, I made big eaves in the building to create more shade. I enclosed a minimum area with fittings because

I thought that fittings create a vague boundary between the inside and the outside of the house, especially when one opens and shuts the doors. I suspended the 2nd floor from the roof by using pillars of thin steel and did not use any pillars at the 1st floor, so as to eliminate the boundary between the inside and the outside of the house even further. The pillars give rhythm to the space of the 2nd floor. I used the colour orange on the outside wall so that the house can exist as part of the big landscape. With the usage of orange, a complementary color to the blueness of the sky, I aim to produce a synergy effect.

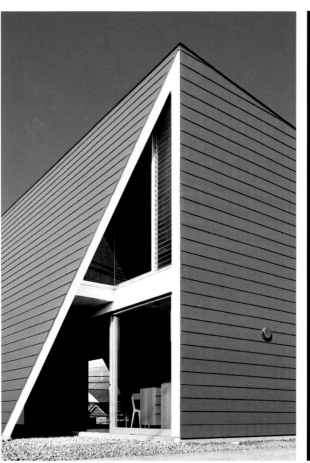

1. Kitchen
2. Dining room
3. Living room
4. Bath room

First floor plan

Section

内面外向式住宅 Inside Out

Location Tokyo, Japan
Site area 150㎡
Bldg. area 59㎡
Total floor area 91㎡
Architecture design Takeshi Hosaka Architects / Takeshi Hosaka
Photographer Koji Fujii (Nacasa & Petners Inc.)

Inside the volume, there is a space in which you would feel yourself being outside: light, wind or rain could enter in, so that your way of living depends on the weather; but in this house, you would always positively seek to find another images of life. Areas on the floor on which rain falls vary according to the wind direction, so you would seek to find the area where you don't get wet. As you live longer, you would find out, from your experiences, various things about the relation between the extent to which rain enters in on the one hand, and the location of objects, furniture and yourself on the other. And, there is no air-conditioner in this house. During intermediate seasons or summer, both humans and cats live in natural draft, or, in other words, they live in the air environment which is almost the same as the outside. On windy or chilly days, you could stay in the indoor boxes or the living room with glass-sliding doors closed. Figures and other many items which had been collected are located not only indoors, but many of which are also put on the outdoor shelves to the extent that is possible. Cats walk on the thin line on which rain doesn't fall and find places in the sun to take a nap. The couple stay in the living room upstairs with glass doors open, sometimes even on rainy days; they often live a life in which they feel themselves being outside even while being inside.

People living in the modern era attempt in the modern way to reduce energy consumption and to coexist with nature - this is

also a positive attempt to find what ways of life are possible beyond energy problems and this, in turn, beyond way of life, leads to the important themes about human mental activities. The couple and cats who had lived in an apartment are now pioneering the new images of life every day in the weather-dependent house inside which they feel being outside.

1. Living-dining
2. Bedroom
3. Toilet
4. Cat window

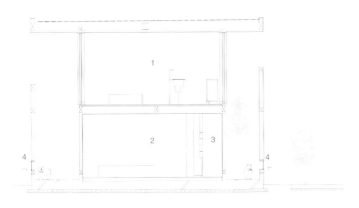

Section 1

1. Living-dining
2. Bedroom
3. Laundry
4. Terrace

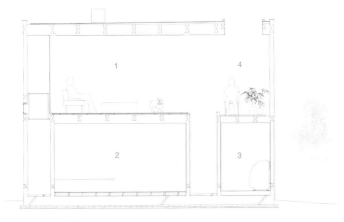

Section

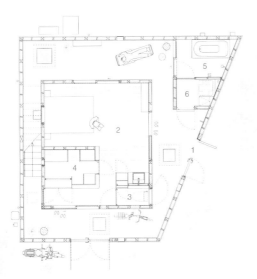

First floor plan

1. Entrance
2. Bedroom
3. Toilet
4. Closet
5. Bathroom
6. Laundry
7. Living-dining
8. Terrace

Second floor plan

鸟巢住宅 Bird House

Location Nagoya, Japan
Site area 428.21㎡
Built area 113.74㎡
Total floor area 117.46㎡
(Main building / First floor area :
40.27㎡、Second floor area : 67.89㎡
Annex / First floor area : 9.30㎡)
Architecture design Katsuhiro
Miyamoto & Associates
Principal in charge Katsuhiro
Miyamoto
Design participation Keishi Yamamoto,
Takenori Uotani
Structural engineering Masaichi
Taguchi / TAPS
General contractor IDO KENSETSU
Photography Katsuhiro Miyamoto &
Associates

It is difficult to place architecture in an inclined, raw site, in "nature". Earthwork is always needed. In the case of building on a steep site, usually a retaining wall is built and sites will be prepared in tiers. In other words, the earthwork plays a role as a kind of bridging of architecture and nature. In that case, there aren't any other* ways to adapt artifice to nature more gently by a structure which is something between earthwork and architecture, a little more elegant than a retaining wall. I think earthwork is, firstly, more infrastructural than architectural. What I wish to pursue is alternative potentialities of earthwork as light infrastructure.

The lobster-claw-like foundation for the "Bird House" hooks the house to the landscape and is the result of our study for the potential of earthwork as light infrastructure. Rather than leveling the slope, you can catch the slope with spikes and still guarantee to build securely. This is the role for earthwork which was originally required in the relationship between architecture.

By making use of the site characteristics, adjacent to the roads at the top and bottom of the site, RC foundation forms a zigzag approach just like a mountain trail to connect these two roads. This is exactly what is called infrastructure. Before the building is built, the site is barrier-free and this RC foundation provides the access for construction. And at the landings formed at the turning points of the zigzag approach, there are three "sites". The name "Bird House" was given to the three cute white houses nested on the branch-like foundation.

1. Private Room
2. Closet
3. Anteroom
4. Storage
5. Entrance 01
6. Porch 01
7. Porch 02
8. Annex-Tatami Room

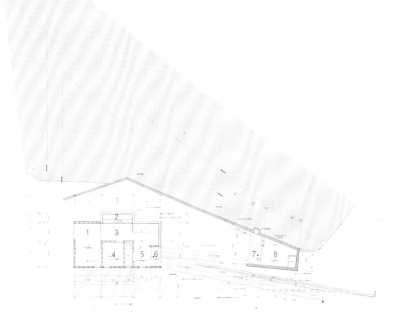

First floor plan

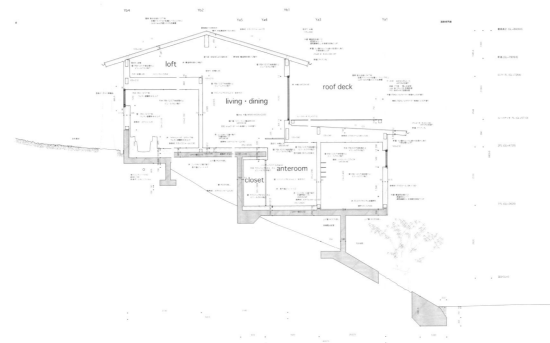

loft

living · dining

roof deck

anteroom

closet

Section

Gapyeong住宅 **Gapyeong House**

Location Gyeonggi-do, Korea
Site area 1,372㎡
Blgd. area 99㎡
Total floor area 128㎡
Architecture design Noble
Construction Luxury Construction
Design Group / Ryu Myeong
Design participation Lee Dong-jin

I had intended to design a house that looks like a gallery instead of an ordinary house. A house with good natural lighting had been designed. It was quite extraordinary in terms of good harmony between its layout plan and the elevation scheme. A corridor is provided between the living room and the kitchen. It gives independency to each room and enhances the movement-line. Both of the external and internal spaces look much bigger due to the attic and open-ceiling, which had been made by utilizing the space in the roof.

1. Entrance
2. Living room
3. Room
4. Kitchen/Dining room
5. Terrace
6. Storage

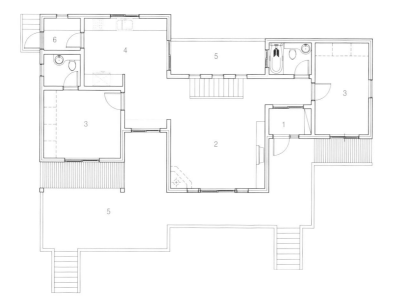

First floor plan

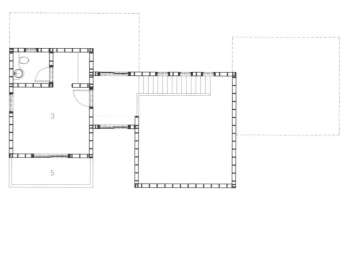

Second floor plan

Front elevation

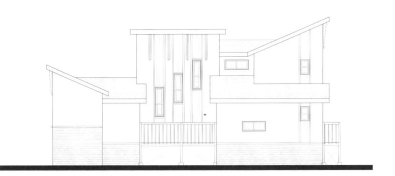

Rear elevation

花园最大化住宅 The Maximum Garden House

Location Jalan Rendang, East Coast,
Singapore
Site area 350㎡
Total floor area 340㎡
Architecture design Alan Tay,
Formwerkz Architects

Design approach

We are interested in the notion of "landed-ness" in the terrace and semi-detached landed typology. With regard to these types of real estates, the building footprint often occupies most of the plot, leaving little room for outdoor space. However, these outdoor spaces are essential for one's connection with the land or at least the perception of living in a landed property as opposed to living in a high-rise condominium. The semi-detached house at Jalan Rendang gave us the platform to investigate ways to expand this idea of "landed-ness" when the building usually occupies the maximum allowable footprint. Our key strategy was to seek out and reclaim incidental spaces or surfaces of the building envelope. The main areas we look into include the front boundary wall, the car porch and apron roof, façade and the main house roof. If these incidental spaces or surfaces can be reclaimed, we potentially gain back 100% of the outdoor space lost to the building.

Maximum garden

"A garden is a planned space, usually outdoors, set aside for the display, cultivation, and enjoyment of plants and other forms of nature."

We intend to expand the notion of "landed-ness" by creating as many gardens in these incidental spaces. In the process of greening, we began to look into the relationship of garden and architecture as well. The planter screen we created on the

façade that shields the master ensuite from the street was a design that, in our opinion, began to blur the boundary between the two disciplines. The curtain of plants coincides building performance with man's affinity for nature.

Taking advantage of the staggered section of the house, we created a sloping roof terrace that retained a continuous flow from the indoor. The sloping roof-scape reminds us of an undulating terrain. We imagined the inclined plane to be more conducive to sit or lie down and have a conversation while looking out in the same direction, sharing the same moment, like one does in a park. The generous outdoor space created on the roof is ideal for outdoor dining and barbeque, offering an unparallel vantage point of the neighborhood.

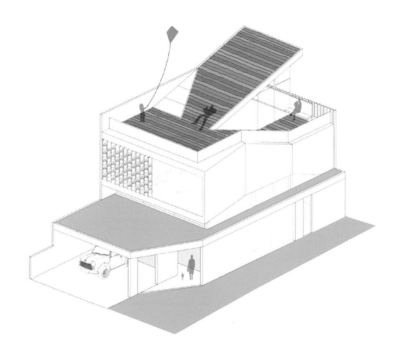

Axonometric

1. Carporch
2. Powder Room
3. Entrance Foyer
4. Living
5. Dining
6. Kitchen
7. Apron
8. Household Shelter
9. Laundry
10. Gusst Room 1
11. Guest Bathroom 1
12. Water Closet
13. Helper's Room
14. Roof Garden
15. Void
16. Study
17. Family Room
18. Boys' Room 1
19. Boys' Bathroom
20. Boys' Room 2
21. Master Bathroom
22. Master Room
23. Gusst Room 2
24. Guest Bathroom 2
25. Roof Terrace
26. Sunken Lounge
27. Skylight

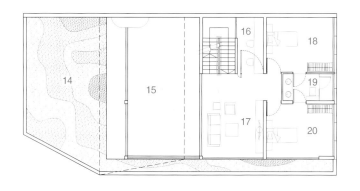

Second floor plan

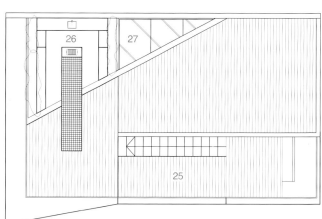

Roof floor plan

First floor plan

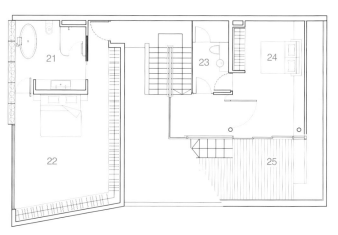

Attic floor plan

Sectional perspective

公寓APARTMENT

卡拉班切尔公寓 Carabanchel Housing

Location Madrid, Spain
Programme 52 single dwellings, 35 double dwellings, 15 triple dwellings + 104 parking slots + commercial space
Site area 4,446.15㎡
Total floor area 8,382.15㎡
Stories 7 (6 + basement)
Built area 12,277.15㎡
Architecture design dosmasuno arquitectos – Ignacio Borrego, Néstor Montenegro and Lina Toro
Photographer Miguel de Guzmán, Alberto Nevado, dosmasuno

Footprint - Orientation and introspectiveness

Despite the guidelines drawn on the plots, places need to express their own personality, to arise naturally, to construct themselves. And concretely this one is aligned against a green area, against the concatenation of public spaces that link the old Carabanchel district with its forest through the new neighbourhood. In response to these conditions, the dwellings are compressed onto one edge, onto a single linear piece, in search for the genus loci of the place, views and an optimal orientation in which east and west share the south, generating the limit of the activity, soothing the interior and defining the exterior.

Strategies - Minimum core + additions

The dwellings are designed from an invariable core with a modulated addition which completes the requirements of the program. This fixed core is constructed attending to the surrounding views and sunlight, and its two main pieces, living and sleeping rooms, are stacked to the south limit, from which they are protected with a filter, relegating a services' strip to the back side. Behind this strip, and like clouds drifting over the void, variations are introduced by the addition of programmatic pieces that form the dwellings of two or three bedrooms. The strict order achieved by the linear core is mathematically blurred into a shifting volume. Thus, the dwellings become "machines for living", and they are designed as such, fitting surfaces and diminishing the transitional areas of between rooms.

Built - Modular casting system

Its construction responds to a necessity of industrial optimization. Therefore, the structure of the main body is constructed in concrete from a single high accuracy aluminium cast. At the same time, the light steel structure modules that constitute the additioned elements enable volumetric variations. This industrialized system facilitates the constructive process, avoiding rubbish and accelerating the implementation times.

South elevation

North elevation

Sección Este

Section (East)

East elevation

Floor plan 1

Floor plan 2

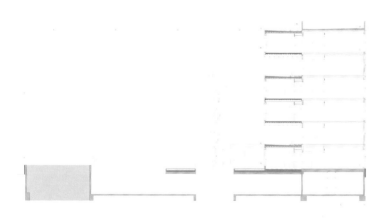

Section detail 1

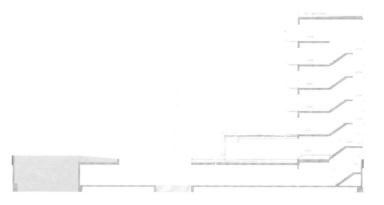

Section detail 2

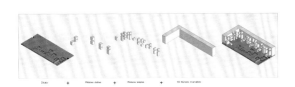

Development scheme

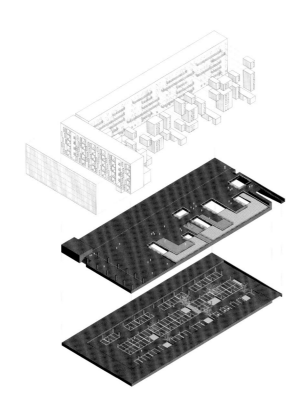

Axonometric

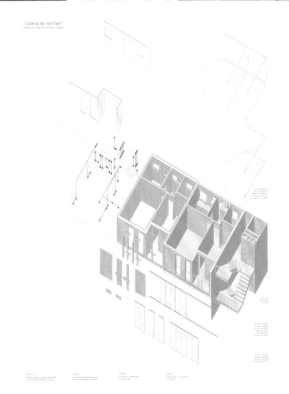

Construction process

Cerredo群体住宅 Cerredo Social Dwellings

Location Degaña, Asturias, Spain
Gross internal floor area 2,385 m^2
Architecture design ZON-E
ARCHITECTS, Nacho Ruiz Allén &
José Antonio Ruiz Esquiroz
Design participation Lucía Martinez Trejo,
Sara López Arraiza
Photographer Ignacio Martinez, Jose
Antonio Ruiz
Client Principado de Asturias Government

This project comes up from a tendering process to build state subsidized housing in Cerredo (Asturias), a mining town located in the very heart of the Cantabrian Mountains where no residential construction had been made for over 25 years.

The project has two stages that materialize in two perpendicular buildings forming an L. In the first stage we undertake the biggest building, which faces the road that crosses the town.

The volumetric we propose has an angular shape. It is a geometry crystallized from some elemmentary laws that are given by the town-planning regulations. The formal result is something halfway between a petrified object, a mountain's shape and a disturbing organism floating over the mountainside.

This "crystallographic" object has the same dark color as the local slate. Like a piece of coal, it absorbs almost all the light it gets and reflects a small amount of it, calmly showing us its rich geometry.

The building's unity contrasts with the individuality of each of the 15 apartments that show through some galleries in the facade. These are cubes which drill the volume using a herringbone pattern and work as heat and light exchangers.

Each of the apartments is different, both in size and in its floor plan distribution, in the location of its gallery and in its roof's configuration. However, all of them enjoy cross ventilation and breathtaking views of Asturias' craggy landscape.

The project's nature as object is emphasized by the way the groud floor is approached: this has been set back along its perimeter, reinforcing the idea of a "floating body."

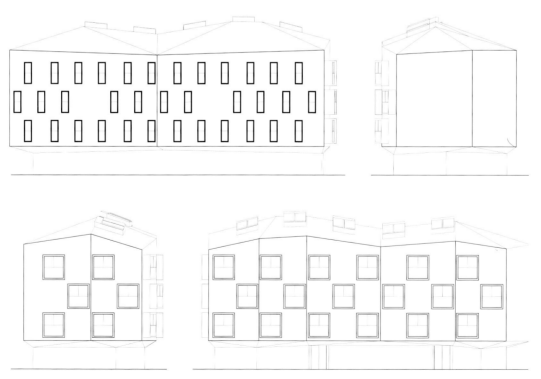

Elevation

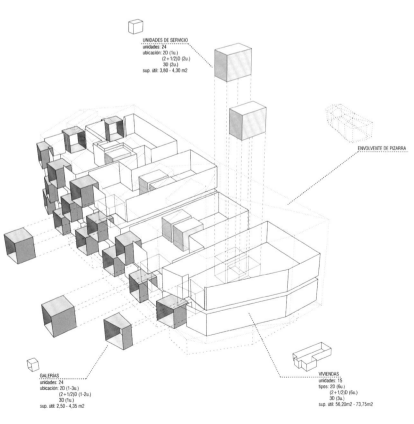

UNIDADES DE SERVICIO
unidades: 24
ubicación: 2D (1u.)
(2 + 1/2)D (2u.)
3D (2u.)
sup. útil: 3,80 - 4,30 m2

ENVOLVENTE DE PIZARRA

GALERÍAS
unidades: 24
ubicación: 2D (1-3u.)
(2 + 1/2)D (1-2u.)
3D (1u.)
sup. útil: 2,50 - 4,35 m2

VIVIENDAS
unidades: 15
tipos: 2D (6u.)
(2 + 1/2)D (6u.)
3D (3u.)
sup. útil: 56,20m2 - 73,75m2

Axonometric

邦代海滩住宅 The Bondi

Location Bondi Beach NSW Australia
Site area 1,284.45㎡
GFA 5,997.9㎡
FSR 4.67:1
Typical floor plan area 521㎡, 1,047㎡
Architecture design PTW Architects
Project director Andrew Andersons AO
Principal Director,
PTW Architects
Photography Sharrin Rees,
Andrew Andersons, PTW Architects

PTW Architects' winning competition design involved the conversion of an existing nine storey motel into a superlative contemporary apartment building with ground floor as retail and restaurants.

Reminiscent of the elegance of 1930s-era architecture, The Bondi's glass curves and striking angles across its exterior create a bold statement against the skyline, responding to both the beachfront location and the unique architectural flavour of Bondi's iconic esplanade.

The Bondi features two striking design components: at the lower levels, the "podium" is wrapped with art-deco inspired bay windows, to mirror the facades along Bondi's promenade, while the dramatic curves of the "tower" levels, standing high above the surrounding streetscape, define its majestic external form. The two storey podium provides studio and one-bed apartments. To complete the variety of accommodation the three-bed apartments in the tower have glazed wintergardens and bays overlooking the ocean.

The building has the highest possible standards of environmental measures incorporated in the design to conserve energy, prevent water and air pollution and manage waste produced by the building. These include gas-boosted solar water heating panels, rainwater harvesting with a 5,000 litre water tank and

@ Sharrin Rees

recycling of over 50% of the existing structure. Construction of the Bondi commenced in June 2007 and was completed in May 2009.

The building has been designed to the highest design standards, with subtle contemporary architecture recalling the style of early modern and art deco era buildings and allowing the Bondi to sit comfortably amid the elegant 1930's-era architecture that curves the length of the beachfront promenade.

@ PTW Architects

@ Andrew Anderson

@ PTW Architects

@ Sharrin Rees

Elevation

@ Sharrin Rees

@ Sharrin Rees

@ Sharrin Rees

Villiot Râpée住宅 **Villiot Râpée**

Location Paris, France
Gross floor area 5,120㎡
Net floor area 4,002㎡
Architecture design Hamonic + Masson
(Gaëlle Hamonic + Jean-Christophe
Masson)
Photography Grazia, Delangle
Program 62 council flats + assiciation's
office
Projet manager Marie-agnès de
Bailliencourt

A breath of fresh air

The project embraces new concepts of living together primarily based on generous outdoor spaces, both private (balconies) and communal (floor area), as well as on an extrapolation of the advantages of detached houses, which have now disappeared forever from Paris – having one's own floor space and thus being rooted in the soil.

The project's answer to such a high density of buildings is to be dynamic, go aerial and disrupt the status quo. A new mood is sweeping through landscape and houses, strong enough to lift them off the ground and send them spiralling up into the air.

It is this link to the ground that brings coherence to the project. It can be felt on every level and in all aspects of the building, with the green balconies (quite thick, like a floating mass) also accentuating this impression that the ground has taken off. This stacking of floor levels defines the architecture of the buildings and the public spaces. The starting point of the project, the ground, accompanies and moulds itself to the natural level, twists and transforms itself, hosts different disciplines, guides and accompanies residents, visitors and passers-by and makes it into a home.

Each level and each flat has a different floor lending itself to different practices and uses. Rather than being like a balcony, a loggia (or a terrace), which can be seen and used on a daily basis, winds its way around the outside of the flats and gives residents the feeling that they live outdoors. This "poured garden" creates close ties to the building's external environment.

Two hybrids

The project involves two blacks of flats, one of 11 storeys above the ground floor and the other of 8 storeys above ground floor.

© Grazia

© Grazia

They stand out not only because of their height but also because of their movement, one being a hybrid of the other and their proximity creating the impression of shifting morphology. They are connected by their moving relationship to the ground.

A path crosses the block in an arc running from the entrance on rue Villiot to the fire service access on the quai de la Rapée. This walkway, which is covered in a green-coloured soft material, is bordered by a garden filled with trees and plants.

Eyes wide shut

Climate planning and sound-proofing have also left their mark... and permitted a system of "truly outdoor spaces" that are therefore independent of the internal floor-plan, creating a stack of more or less closed terraces and more or less open loggias wrapped around the four sides of each tower like a "serpentine". One can stroll around a flat, walking out of the bedroom and into the living room; there are many paths to choose from and many surprises in store. But the main thing is that no one can see their downstairs neighbour, and the flats opposite are occluded by screen walls and balustrades.

Inside, the flats are arranged around a central structural core that houses the flows, stairs and lifts. Each landing serves three to four flats. Only these cores and the façades are weight-bearing, which means that the decks can be opened up and the floor-plan reversed. Flats today — or tomorrow?

© Delangle

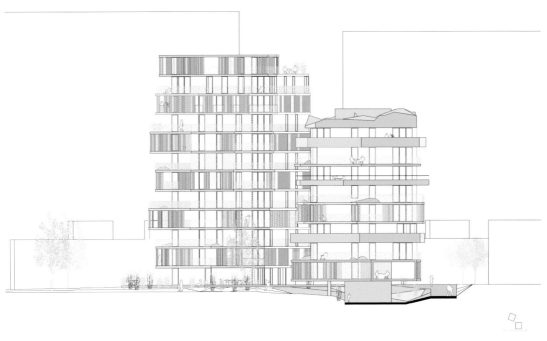

Section east

© Grazia

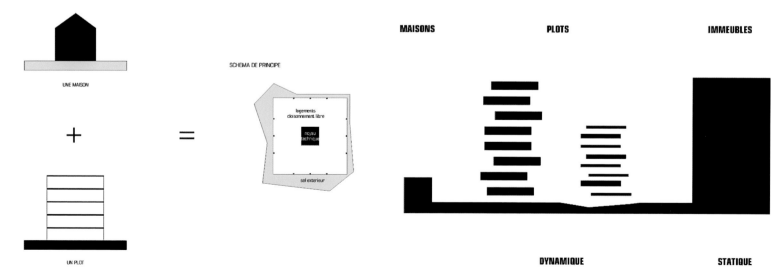

UNE MAISON

SCHEMA DE PRINCIPE

MAISONS   PLOTS   IMMEUBLES

+

=

logements
cloisonnement libre

noyau
technique

sol exterieur

UN PLOT

DYNAMIQUE   STATIQUE

Concept

混合住宅 Linked Hybrid

Location Beijing, China
Total floor area 221,462㎡
Architecture design Steven Holl
Architects / Steven Holl
Interior design Steven Holl Architects
/ China National decoration Co., LTD
Photographer Shu He, Iwan Baan, SHA

The 220,000㎡ Linked Hybrid complex in Beijing, aims to counter the current privatized urban developments in China by creating a twenty-first century porous urban space, inviting and opening to the public from every side. A filmic urban experience of space: around, over and through multifaceted spatial layers, as well as the many passages through the project, making the Linked Hybrid an "open city within a city". The project promotes interactive relations and encourages encounters in the public spaces that vary from commercial, residential, and educational to recreational, a three-dimensional public urban space.

Focused on the experience of passage of the body through space, the towers are organized to take movement, timing and sequencing into consideration. The point of view changes with a slight ramp up, a slow right turn. The encircled towers express a collective aspiration rather than towers as isolated objects or private islands in an increasingly privatized city, our hope is for new "Z" dimensional urban sectors that aspire to individuation in urban living while shaping public space.

Geo-thermal wells (655 at 100m deep) provide Linked Hybrid with cooling in summer and heating in winter, and make Linked Hybrid one of the largest green residential projects. The large urban space in the center of the project is activated by a greywater recycling pond with water lilies and grasses in which the cinematheque and the hotel appear to float. In the winter the

pool freezes to become an ice-skating rink. The cinematheque is not only a gathering venue but also a visual focus to the area. The cinematheque architecture floats on its reflection in the shallow pond, and projections on its facades indicate films playing within. The first floor of the building, with views over the landscape, is left open to the community. The polychrome of Chinese Buddhist architecture inspires a chromatic dimension. The undersides of the bridges and cantilevered portions are colored membranes that glow with projected nightlight and the window jambs have been colored by chance operations based on the "Book of Changes" with colors found in ancient temples.

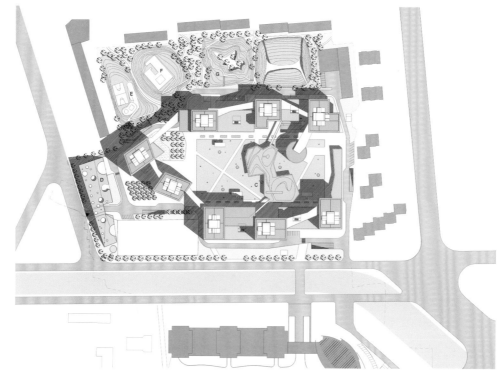

Site plan

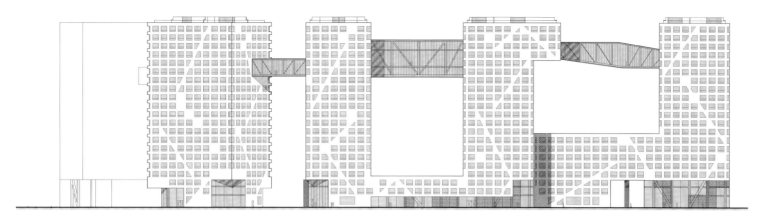

Elevation

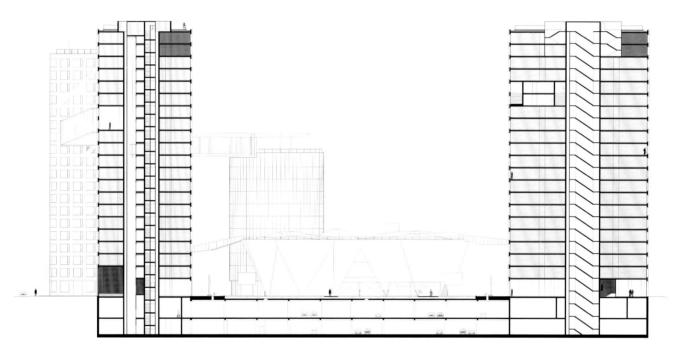

Section I

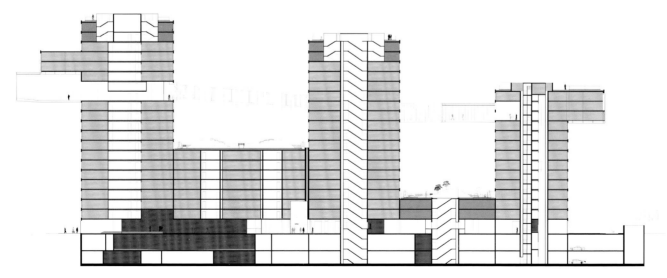

Section II

First floor plan

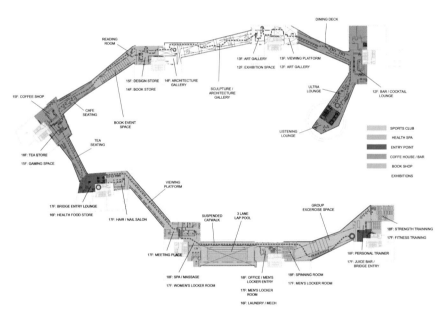

Diagram (program)

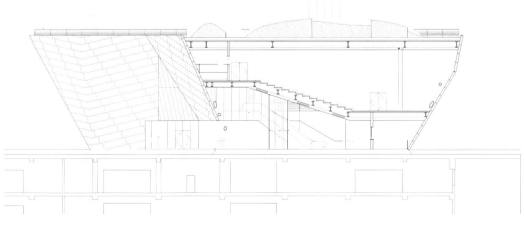

Detail (theater)

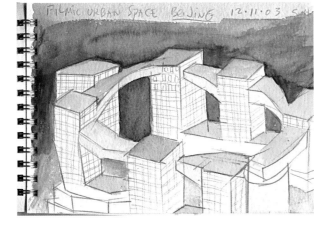

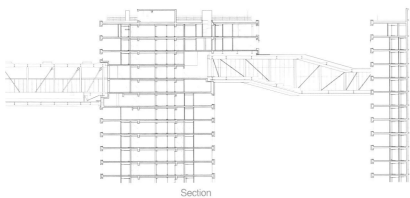

Section

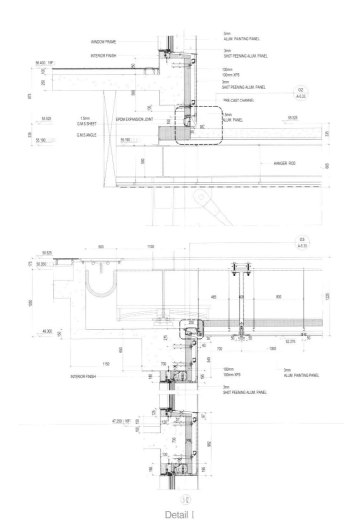

Detail I

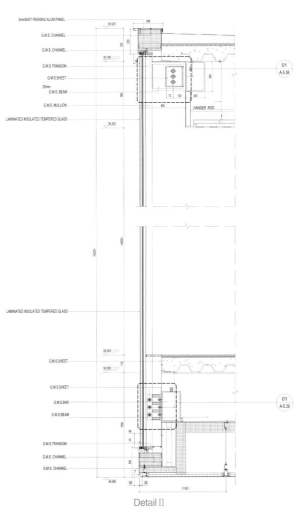

Detail II

侧边住宅 Siloetten

Location Aarhus, Denmark
Total floor area 3,000㎡(Housing),
1,500㎡(Village center)
Architecture design C. F. Møller
Architects + Christian Carlsen
Arkitektfirma
Design participation C. F. Møller
Architects in collaboration with
Christian Carlsen Arkitektfirma
Interior design C. F. Møller Architects
Photographer Julian Weyer

Many towns in Denmark have centrally located in industrial silos; most are no longer in use, but continue to visually dominate the local skyline. This is also the case in the town of Løgten north of Aarhus, where the former silo complex has been transformed into a "rural high-rise", with 21 high-quality residences composed as individual and unique "stacked villas".

They are an alternative to standard apartments or to detached suburban sprawl, and are a mix of single storey flats and maisonettes, meaning that even the lower levels fully get to enjoy the views, and that no two flats are the same. The actual silo contains staircases and lifts, and provides the base of a common roof terrace. Around the tower, the apartments are

built up upon a steel structure in eye-catching forms which protrude out into the light and the landscape — a bit like Lego bricks. This unusual structure with its protrusions and displacements provides all of the apartments with generous outdoor spaces, and views of Aarhus Bay and the city itself. Similarly, every apartment enjoys sunlight in the morning, mid-day and evening, whether placed to the north or south of the silo structure. At the foot of the silo, a new "village center" is created, with a public space surrounded by a mix-use complex with shops, supermarket and terraced housing, and a green park containing small allotments for the residents.

The nature of the silo's "rural high-rise" remains unique —

since it is a conversion, no other building in the area can be built to the same height, and it will remain a free-standing landmark. It is an example of how the transformation of redundant structures can hold the potential both to give a new identity, and to introduce density to suburban outskirts. The body of the silo is deliberately left visible on the side of the building facing the new center, to ensure a continued legibility of the history of the site, and to acknowledge that these types of structures have an equal validity as rural historical markers, for instance the church bell-tower or historic windmills.

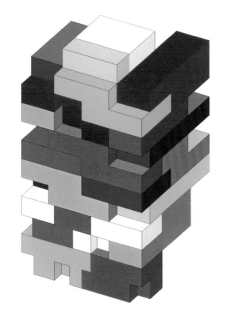

Sketch

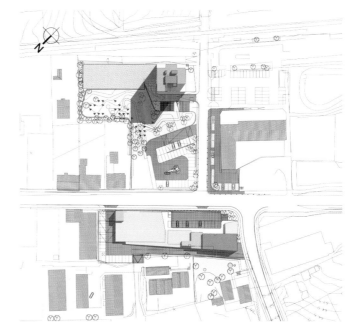

Site plan

North elevation

West elevation

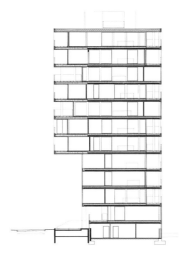

North section_units

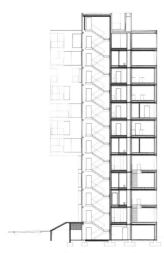

North section_staircore

Level 4

Level 11

Level 2

Level 6

Level 1

Level 5

波士顿大学学生公寓
Boston University Student Housing

Location Sydney, Australia
Architecture design Tony Owen
Partners and Silvester Fuller Architects
Concept design Silvester Fuller
Photographer Brett Boardman

The new student quarters for Boston University by Tony Owen Partners and Silvester Fuller Architects at 15-25 Regent Street, Chippendale is a unique design using fissures to provide maximum solar access to bedrooms as well as natural ventilation throughout the building.

The eight-level, environmentally-efficient building has been designed to house overseas students visiting Boston University for a semester in Sydney. It can contain 164 beds as well as a care-taker suite. It also has three lecture halls, a library, an internet lounge, a rooftop terrace with a timber deck and an adjoining fully-equipped communal kitchen, plus cafe. The unique module houses 4 single bedrooms for every shared lounge and bathrooms.

The concept design was by Silvester Fuller and was developed through subsequent approvals, documented and completed by Tony Owen Partners. Construction was by Ceerose Pty Ltd.

Arup Partners provided the environmental analysis. The design uses large canyon-like slots in the façade which allow sunlight and ventilation to penetrate deep into the building and into each room.

The windows in these slots have a rhomboid shape to maximize efficiency, and deliver a bold architectural façade which is illuminated at night through an ever-changing light show. The windows are oriented to trap sunlight whilst ensuring privacy between rooms.

The end walls of the slots are made from glass louvers that are seven storeys high, and the building also contains a seven-storey glass louvered atrium. Air is drawn through the canyons and passes through the building like gills. It is drawn up through the central void and out of the top to ventilate corridor areas, thus allowing the building to breathe naturally.

East-facing operable louvers on each level further help to lower ambient temperatures by drawing in fresh breezes.
The design allows more light and ventilation into each bedroom, provides good views, and is a sensible model for residential density in the CBD.

At the same time the design is a sensitive infill for a constrained

urban site. Regent Street contains a number of heritage buildings. The facades utilize masonry to create a contemporary contextual response that is still contemporary. Colour is used on the ground floor structure to celebrate the student life in the street. The corner of the building is set back in the northern corner to widen the lane way. A café and outdoor seating is located here to enliven the lane way and create an adhoc public space.

At night the fissures are lit up at night. This will be programmed with changing colours to create a permanent light installation. A 7-storey light installation called "fluid dynamic" is being installed in the central atrium.

A large central stair fissure directs visitors to the below ground theatre and lounge spaces. This fissure has a fractal roof and as well as being a ceremonial entry, helps draw light down into the building. The fractal roof is repeated in orange perspex in the café.

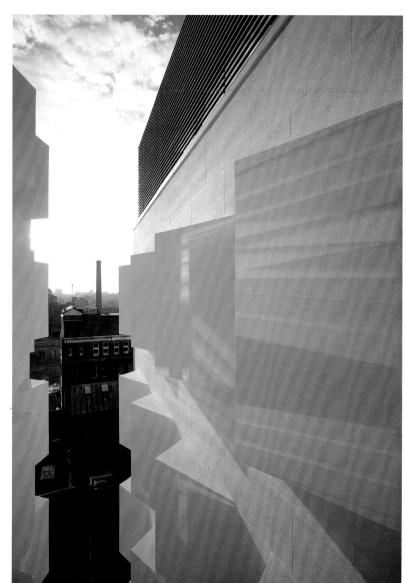

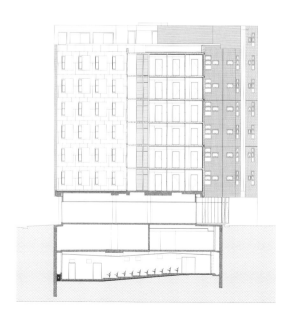

Section

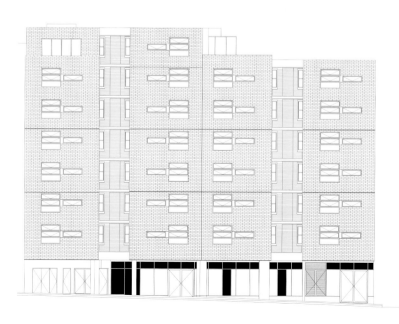

East elevation

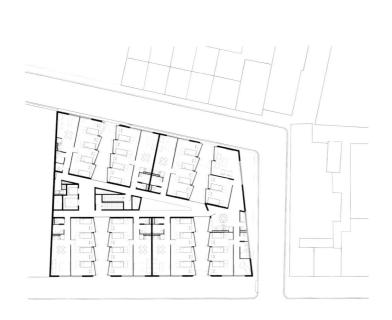

Typical plan

Living Foz住宅 Living Foz

Location Porto, Portugal
Built area 13.700 sqm + courtyard

Architecture design dEMM
arquitectura, Paulo Fernandes Silva

Landscape design arqt.OF
Photographer Pedro Lobo, Fernando
Guerra
Collaborators Isabela Neves, Tiago
Soares Lopes
Client J. Camilo

Living Foz is located in Foz do Douro, an Oporto city parish with ancestral occupation due to its south sun exposure and proximity to both the sea shore and the Douro River mouth. Nowadays, it struggles to consolidate and qualify its residential offer. Therefore, Living Foz was expected to present a new approach to high quality residential buildings, emphasizing the relationship between interior and exterior, taking advantage of the sea shore views and the portuguese sunny climate.

Accomplishing these expectations, the balcony angle articulation creates outdoor spaces that are enriched by contrasts of light and shade, exposure and protection. The shape of the balconies is emphasized by the contrast between the white cast-in-place concrete and the dark glass reinforced concrete panels.

The interior compartments, articulated in order to encourage a flexible use of the social areas, are widely opened to the exterior by large roof to ceiling sliding windows. Nevertheless, these glass surfaces are always protected from direct sun exposure, representing a simple though effective way to reduce energy consumption. Unleashing 30 percent of the plot area for a collective use garden, a relief to the neighboring construction is obtained, qualifying the surrounding residential area. The geometric and material principle of the building continues through the external areas, defining the direction and shape of the garden paths. The species were chosen to create a wide variety of colors and textures, and to expose or dissimulate the public and private spaces of the ground floor.

The 40 apartments (divided in typologies of two to five bedroom) are distributed over 7 floors and completed with private parking in 3 underground floors. The finishing materials are intended to be durable and to reproduce in the interior the contrasts between light and shadow of the exterior. Therefore, the pavement is lined in dark varnished wood, and the walls and carpentry are painted in a very light white tone. Kitchens and toilets are coated with portuguese natural stone.

Living Foz is also about sustainability, regarding the use of high technology insulation in windows, walls and coverture as well as the conception of the building design itself. In fact, the controlled sun exposure, the thermal inertia of the facade, the use of alternative energy and electrically efficient appliances, are features that integrate the building in a quality standard of respect for the individual and their habitat.

© Pedro Lobo

© Fernando Guerra

© Pedro Lobo

© Fernando Guerra

© Pedro Lobo

© Pedro Lobo

© Pedro Lobo

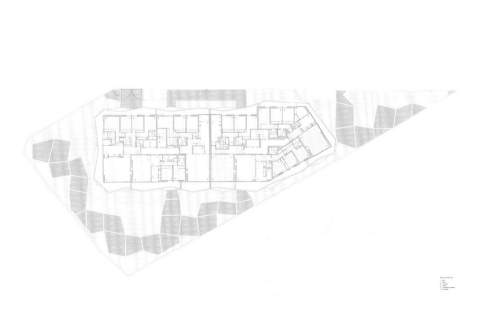

Ground floor plan

© Pedro Lobo

Hans-Jürg Buff公寓住宅
Apartment building Hans-Jürg Buff

Location St. Moritz, Switzerland
Gross area 660m²
Architecture design Pablo Horváth, Architect
Assistance Ferruccio Badolato, Heinz Noti, Gabriela Walder, Andreas Wiedensohler
Site management Peter Maurer, St. Moritz Ralph Grether, Samedan
Client BOKA AG, Hans-Jürg Buff, St. Moritz
Photographer Roger Frei, Zürich

The apartment building Hans-Jürg Buff forms the southwest end of the residential development Chalavus in St. Moritz Bad. Together with the other surrounding buildings, it is grouped around a protected leafy courtyard, comprising a high quality overall urban development. The new five-storey building continues a block, loosely integrated with the existing structure on the south side of the courtyard. The apartment complex benefits from its detached, free-standing setting. The accommodation units are laid out in two-storey units in all four directions, providing stunning views of the picturesque mountain landscape. The rooms fan out following the sun's course from the east side to the south-west. In this way the natural light floods in the living areas and at any time of day there are always variable and interesting light patterns.
The architectural use of forms employed on the new building refer to local construction culture but these traditional and regional elements are conducted into their modern setting.

The imposing figure recalls the typology of residential tower in Graubünden, occurring since mediaeval times. The whole building is cladded entirely in Gauinger travertine and underlined through the use of a irregular bonding in the stone finish it gives a significant association to a overall monolithic structure. This idea is further highlighted in the polygonal shape of the cubage.

Wood-inlaid loggias modulate and add a visual structure to the monolithic edifice in keeping with the building style referenced. The wooden window shutters reflect the character of the region as well as serving contemporary demands.

The insulating layer between the eight-centimetre-thick stone and concrete walls conforms to Canton energy policies, being of a considerably higher heat transfer coefficient than that required. Geothermal probes provide ground-source energy for hot water and heating.

1. Living Room
2. Kitchen
3. Bedroom 01
4. Bath 01
5. Bedroom 02
6. Bath 02
7. Staircase

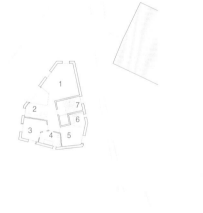

First floor plan

Ground floor plan

Southeast elevation

Section A-A

LeTage住宅 **LeTage**

Location Queensland, Australia
Bldg. area 1,100㎡
Total floor area 170㎡
Architecture design Ellivo Architects
/ Mason Cowle, Scott Whiteoak
Photographer Scott Burrows

Brisbane has rediscovered its river and the inner city flourished with an influx of residents recognising the lifestyle benefits of near city neighbourhoods such as New Farm and Teneriffe as desirable and convenient for work, recreation and amenity. LeTage reflects this social realisation and need for an increased density living offering a preferable option to the large apartment complexes which have become a dominant feature within the inner-city and along the river.

LeTage is evolved from our clients' understanding and recognition of the benefits in providing a high quality living environment which combines the convenience and excitement of inner city living with the space, privacy and amenity of a traditional home. The result

is 5 residences, one per floor, with private lift accesses within a contemporary expression of urban living. The site sits high above the river on the Kangaroo Point cliffs with a commanding view to the Storey Bridge and the city. Internal ponds at the entrance to each residence separate the functional zones of public and private and extend the reference of water as an element connecting the entry gatehouse to the river. A central corridor acts as a breezeway and visually links the resident to the street and the river. The apartment interiors feature timber floors, extensive use of timber veneers, stone and glass in a clean and contemporary palette. Externally, random tiles, render and coral stone express and delineate the levels.

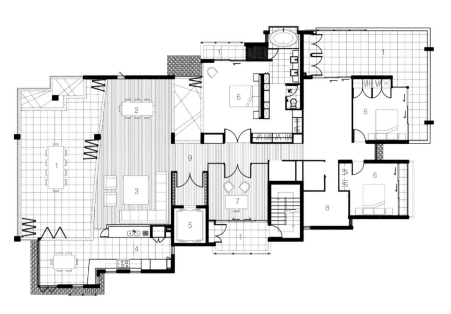

upper floor plan

Elevation I (river)

Elevation II (street)

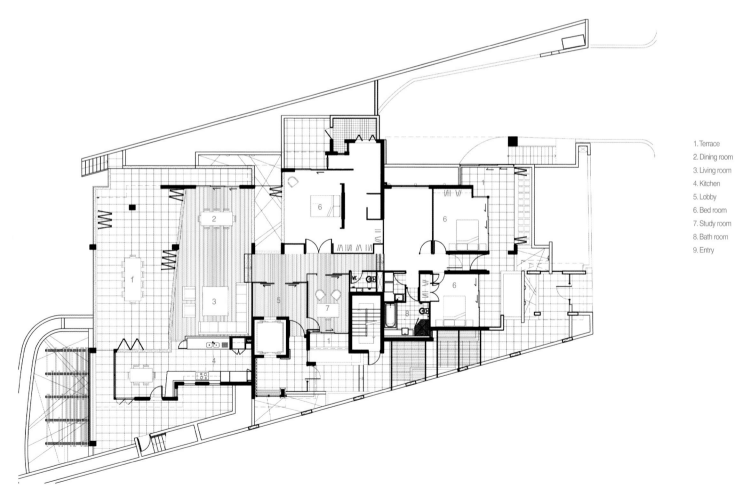

1. Terrace
2. Dining room
3. Living room
4. Kitchen
5. Lobby
6. Bed room
7. Study room
8. Bath room
9. Entry

Ground floor plan

36街区公寓 Block 36 Apartment

Location Westown Cairo, Egypt
Plot area 3,965 m²
BUA 7,000 sqm
Architecture design Shahira H.
Fahmy Architects
Design participation Shahira H.
Fahmy, Jenane Azmi, Laila Badawi,
Mohamed Farid Shaltout, Georges
Talaat, Rafik Hanna
Developer/Client SODIC and
Solidere

Westown is a mixed-use city center located within Sheikh Zayed City which is developing into one of the leading satellite communities of Cairo. Unique placement adjacent to the Cairo Alexandria Highway from the north and cornered by the Dahshour road to the west, West town provides matchless accessibility and exposure. Within West town, Block 36 is nestled at the heart of density and activity. It is the perfect middle-ground located between the quiet purely residential south-east and the pulsing commercial/ recreational north-west. Connected to both the primary pedestrian and vehicular thoroughfares, Block 36 is one of the principal locus blocks from which activity in West-town radiates. The pleasant North wind finds easy access to the blocks North/North-West side

When first approaching this project we thought of the many challenges that face Cairo today, most important of which is the incredibly rapid growth of the city, its population and its infrastructure. This growth has a dual nature; it is in many cases a formal, planned and ordered growth, and yet is sometimes engulfed by an informal, sporadic and chaotic growth. We observed that such a phenomenon was very clearly and visually represented in the Urban language of Westown's surrounding area.

Standing on the site of Westown one can see a wide expanse of desert, yet there is actually a fast moving wave of agricultural land (with a striking linear formality) creeping onto it not two

and a half kilometers away. It is closely followed by a faster moving wave of urban developments and informal dwelling (with a disordered informality) crawling right over the agricultural land.

This superimposition of formality and informality creates a series of solids and voids such as those seen in the sketch above which we used as a guideline for the layout and later drew upon during the development of our elevations and masses.

Superimposition of informal growth over a formal grid

We also drew inspiration from certain architectural patterns that we observed in Egypt and around the world and the needs that they fulfill. Examples of such patterns are the closing of balconies and in-between spaces with screens to fulfill the need for privacy, the creation of overhead sheds and street projections for shade and the erection of boundaries for security.

The eight architectural patterns we drew from and their associated need are: Boundaries for security, Gateways for definition, In-between spaces fulfilling the needs for privacy and shelter from heat and sun, sheds also for the hot weather, projections of building into the street to create shade for retail owners and shoppers, screens also used for privacy and finally stairs and circulation elements that can be perceived from the exterior to provide a certain transparency of function typical in Egyptian farmer's houses.

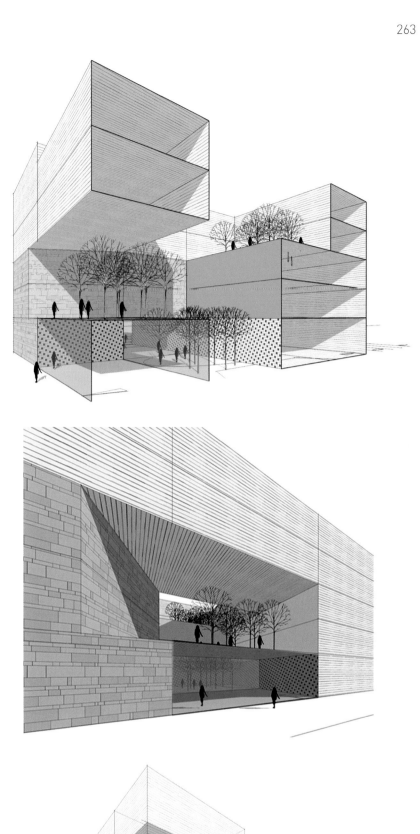

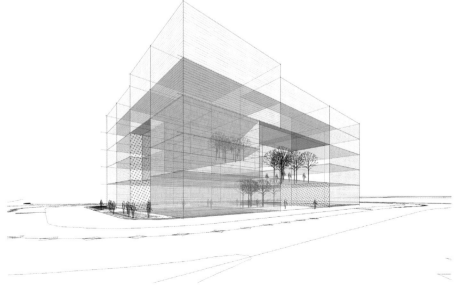

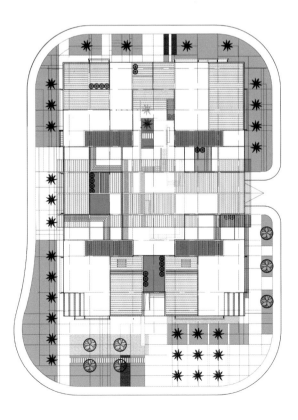

Site plan

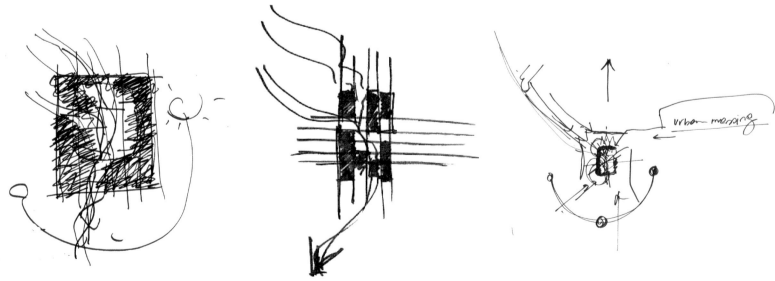

Sketch

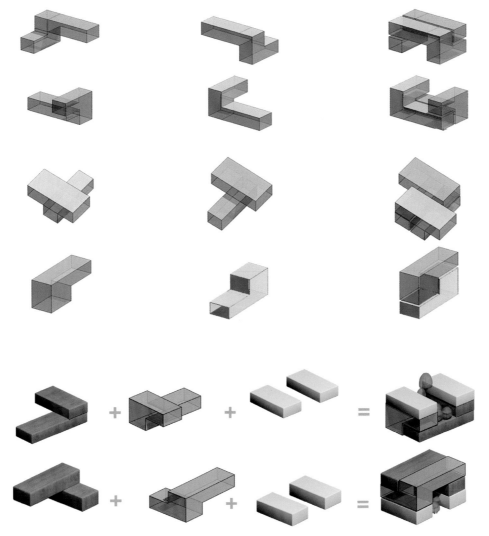

Diagram (Units interlock)

Shuffle住宅 Shuffle House

Location Porto, Portugal
Bldg. area 300㎡
Architecture design ON OFFICE
Design participation Leon Rost,
Ricardo Guedes, Francesco Moncada,
Joao Vieira Costa, Eugenio Cardoso,
Joana Gomes

When Wallpaper asked us to design a house, our immediate impulse was to design an idealistic, conceptual house, on an unrealistic site. Our process quickly reminded us that we needed a real site and a real client with real demands to instruct the design. So we hit the streets to find an overlap between the Wallpaper assignment and a potential project. We finally found a plot, just around the corner from our office, with a client interested in a similar project. We digested the client's programmatic requests and posed the question: "If each room has different demands for space, shouldn't each room claim the space it needs, like a Mondrian grid?"

Shuffle House packs in the variety of spaces within a typical building envelope, while a longitudinal slice of the building designates circulation and services. The facade reinterprets the traditional facades of Porto by playing with scale and rhythm.

■ Process

Intent

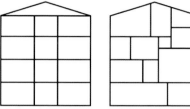

Spaces

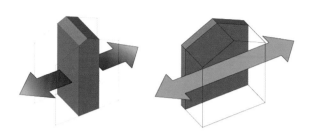

Circulation

■ Facade

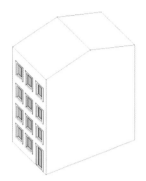

– A typical facade distributes light evenly, regardless of program.

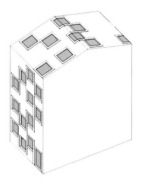

– The Shuffle House facade reorganizes the normal grid of windows to strategic locations to provide light to specific locations within.

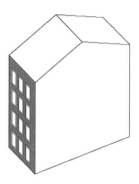

– The traditional Azulejo tiles of Portugal are pieces of art in themselves. When repeated over a facade, the effect from afar is unfortunately reduced to a single tone.

– We propose to celebrate the tile designs by scaling up the single tile design to the width of the building. The design would be composed of individual custom tiles like the historical murals that are preserved in Porto.

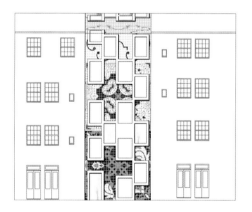

South elevation

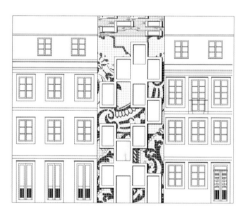

North elevation

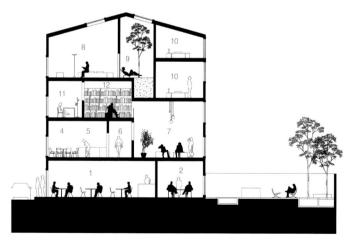

Section I

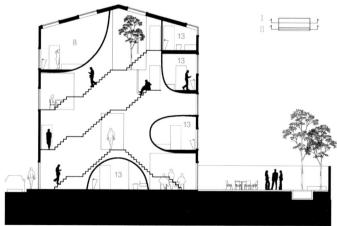

Section II

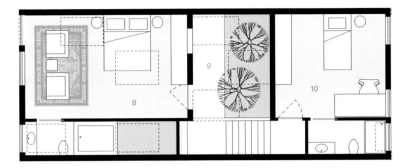

Third floor plan

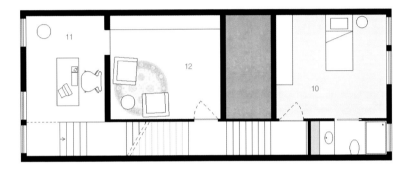

Second floor plan

1. Shop & Cafe
2. Recreation room
3. Backyard
4. Dining room
5. Kitchen
6. Entrance
7. Living room
8. Master bed room
9. Garden
10. Bed room
11. Study room
12. Library
13. Bath room

First floor plan

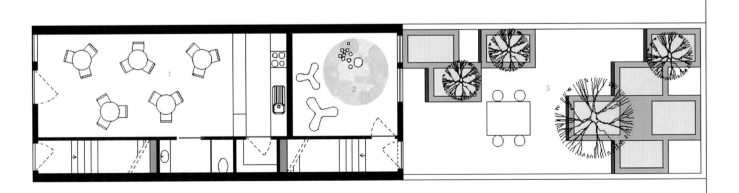

Ground floor plan

立方体住宅 The Cube

Location Beirut, Lebanon
Design Orange Architects,
Netherlands
Surface 3,600m^2
Unit 19 apartments
Height 50m
Client Masharii, Beirut Lebanon

Orange Architects, a partnership between Dutch architecture firms JSA, CIMKA and HofmanDujardin, releases the design of a luxury apartment block on Plot 941 in Sin el Fil, an eastern district of Beirut. The design was commissioned by the Lebanese development corporation Masharii.

The 50-metre-tall block will contain 19 apartments ranging in size from 90 to 180 m^2. The concept for the tower is simple but extraordinarily effective: "maximizing", making optimal use of the site's potential, the local building code and the fantastic views of Beirut and the Mediterranean. Within this conceptual framework, we will develop the most extensive building programme possible. The maximum height will be achieved, and the view from each apartment will be optimised. For this purpose, 100-percent flexible floor plans have been developed. Thanks to the fixed core of lifts and staircases at the heart of the building, there are no constraints on the layout of the apartments. The floors run straight from the core to the exterior walls, which are composed of two transparent façades and two

supporting walls on each floor, rotated 90 degrees per level. The walls, which are perforated, will largely determine the appearance of both the exterior and the interior. The core will also serve to stabilize the tower, which is in a seismologically active area.

From the second floor upwards, each apartment covers an entire floor. The rotation of the volumes on each level will offer residents magnificent outdoor areas on the roof of the apartment below, as well as panoramic windows with views of Beirut. The ground level of the building will hold the lobby, a parking garage for 25 cars (partly set into the adjacent hill), and the standard facilities areas. The roof of the entrance area will serve as an outdoor playground for children.

This project has now reached the final design phase. Last week, both the client and the construction company gave the green light for the next stage. Construction of the tower block is expected to begin in 2012.

Diagram (Concrete Column)

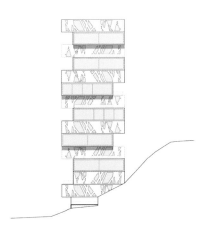

Elevation

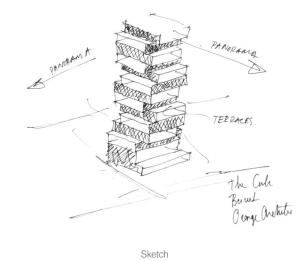

Sketch

941

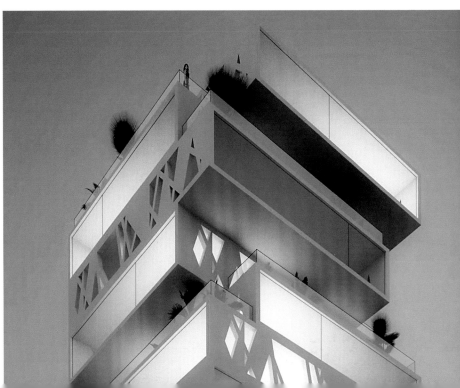

Röntgenstrasse大街的公寓住宅
Apartment House on Röntgenstrasse

Location Zurich, Switzerland
Architecture design
AFGH(Andreas Fuhrimann
Gabrielle Hächler Architekten)
Project leaders Barbara Schaub,
Regula Zwicky

This is a multi-household housing at Röntgenstrasse in Zurich. This 6-storey building is the headquarter of buildings that surround the typical shared-courtyard. On the other hand, it also works as a unique type building.

The main challenge of design was to make the building in harmony with surrounding environment though the project site was a difficult one to locate a building under the Building Code. For instance, a firewall facing the yard can close the surrounding of block when required. The window frames in the facade are different from each other. The cement chimney in the front creates certain dynamic nature in the terrace direction. The chimney makes the building look taller. This dynamic planning gives the existence of this building as an urban architecture. Still, the building essentially has a grid-type facade that reflects the urban context. The design pursued the aesthetics in the modest but clearly visible cement facade. The elegant bronze window, wooden stairway and wooden fall-proof device are in contrast with each other. The building gives the feeling of a handicraft. It has a potential sculptural character together with cubistic three-dimensional nature.

There are two ways for the entrance/exit of this multi-household housing. It is possible to use elevators at the 1st floor entrance. If one would like to use a stairway, there is an open stairway from the courtyard up to the 2nd floor. The 2nd floor and 3rd floor have basic-type households. From 4th floor, one household on each floor has rooms and a dining hall that are open to Röntgenstrasse which gives distant prospect over the railway.

All households have cement walls, cement ceiling, built-in closets and dark-colored oak flooring. Furthermore, all floors have a "balcony room" facing courtyard, which are built different from each other dependent on the taste of each household. This balcony room is in better harmony with the oak flooring than the summer-room open to outside. There are also high-quality balconies in the upper floors that are emphasized by stone flooring and sliding glass-doors.

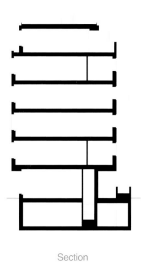

Section

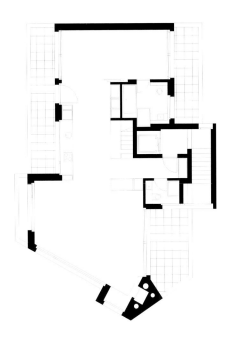

Floor plan 1

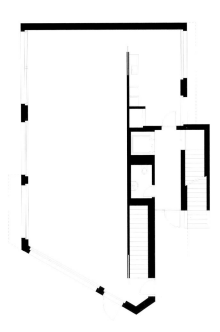

Floor plan 2, 3

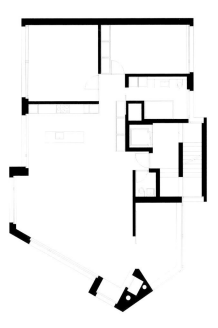

Floor plan 4

ADELHEIM住宅 ADELHEIM

Location Gyeonggi-do, Korea
Site area 4,177㎡
Bldg. area 890㎡
Total floor area 6,095㎡
Architecture design Yoo Kerl
Design participation Lee Ji-eun ,
Ahn Jeong-pyo, Kim Seok-cheon,
Jo Kwang-il

Complex planning

Adelheim is a residential complex that uses shared entrance. The design emphasizes the advantage of independent housing while sharing the economic burden for common facilities, security, maintenance and repair. The project site has good environment with elevation a bit higher than surrounding and sloped. Topography is naturally linked to the Bulgok-san Mountain in northeast. Southwest has a good open view toward nearby residential area. The land belongs to Special Design Zone in the urban planning. The design adopts Ground-connected Town House type. The complex planning emphasizes the advantage of sloped topography and each unit is naturally linked to the green space in the rear. Each unit has privacy like an independent housing and their views are open to southwest, not interfered by the unit in the front.

Units

Each unit in Adelheim has different level from other units. Each unit has its own garden connected to the land giving the sense of independent housing. Unit internal has big and cool space and interior is finished with bright and modern tone. Most of the ground floor space is the living room linked to a terrace facing south with a good view. The kitchen is in the center of ground floor. It is open to the living room so that it would be the center of family life. There is an auxiliary kitchen which can be used dependent on cooking necessity. Master bedroom on the upper

floor is given generous space with an exclusive terrace and sufficient storing spaces. Especially, there is a rooftop terrace directly linked to the master bathroom. It gives the sense of being in the outdoor hot spring.

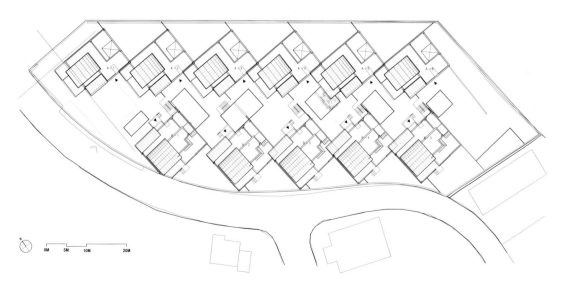

Site plan

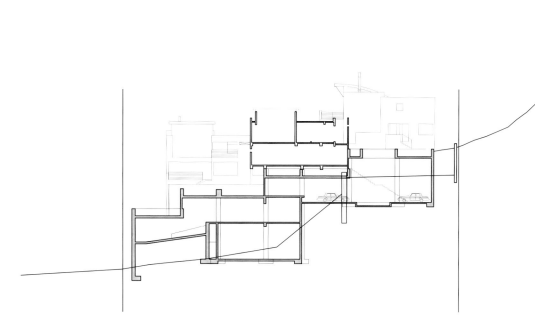

Section

■ Type-A

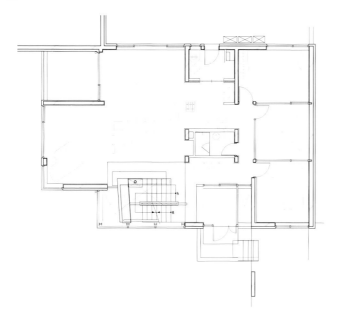

First floor plan

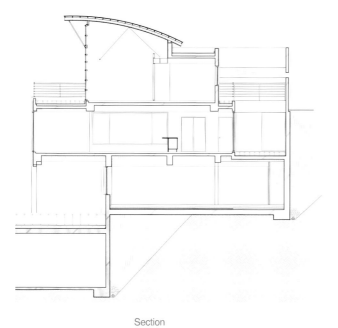

Section

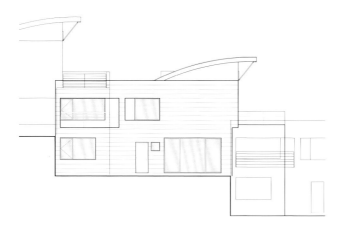

Elevation

■ Type-B

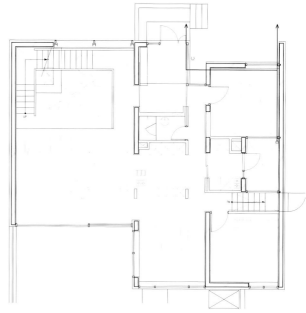

First floor plan

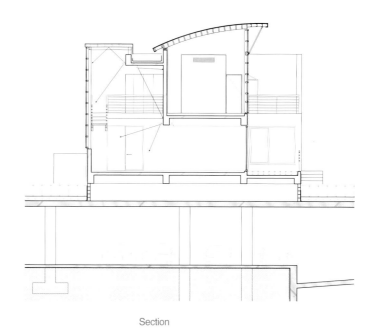

Section

Elevation

Dongtan联排住宅 Dongtan Townhouse

Location Gyeonggi-do, Korea
Site area 10,390㎡
Bldg. area 3,433㎡
Total floor area 7,366㎡
Architecture design Park In-su
Design participation Choi Kyeong-suk,
Hong Sung-kwan, Lee Hyo-yeop, Kim
Doo-jin, Cho Nam-sin

Dongtan Townhouse is designed as a special kind of residential house. We tried hard to achieve advantageous application of building regulation and favorable permit process for the new type of house called an "independent shared house". As the result of our effort, households in Dongtan Townhouse have underground parking spaces of 4 cars per one household and private yard in the front and in the rear. By removing the land boundary between the households, the townhouse has efficient floor planning of 3 aboveground storeys and 1 underground storey. Communities in various types can appear in Dongtan Townhouse because each household will belong to different group through multi-clustering. The shared outdoor space of existing shared houses is owned by everybody but it is not owned by a single one. But the shared outdoor space of Dongtan Townhouse is warranted of exclusive using right since it is owned by a single household. The project site has a slope with elevation difference of 2.0 m. The boundaries of four house rows are naturally made by the slope so that the privacy of outdoor space can be secured in natural way. The roads in the townhouse complex are in the shape of lied "H". These roads are part of the park and the houses are designed as "houses on the park". The houses cannot be attached to each other according to the regulation for independent house. Therefore, the space between the houses is designed as a sunken space. It makes the environment of underground storey more pleasant and minimizes the privacy infringement of aboveground storey

between the houses. Most important thing is the scale of the house. The large scale outdoor space of existing big apartment complex have been just filled with trees and scanty programs. However, the outdoor of Dongtan Townhouse is created by actual use of each household in urban scale. The roads are in random pattern created by houses in various shapes. The roads make the locality of each location and the entrance of each house have different landscape from each other. There are pass-through roads and blind alleys in response to the streetscape of a city.

Site plan

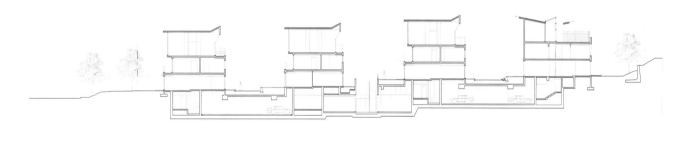

Section I

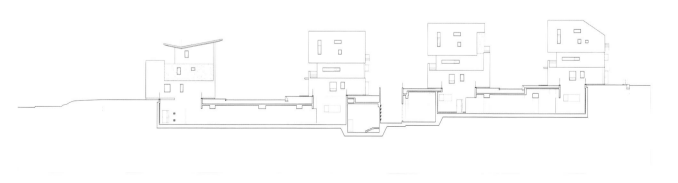

Section II

■ Type-A

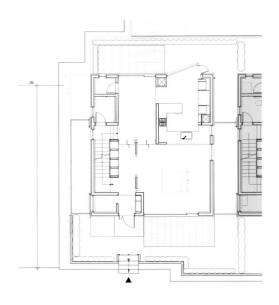

First floor plan

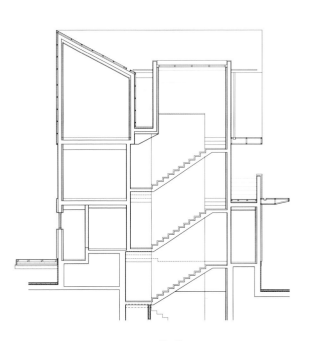

Section

■ Type-B

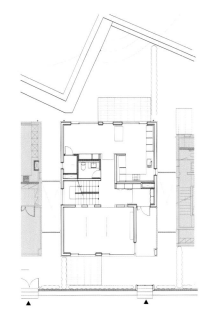

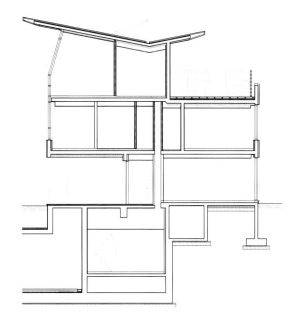

First floor plan

Section

■ Type-C

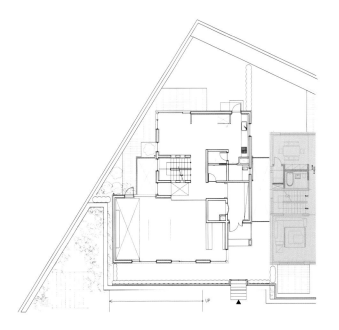

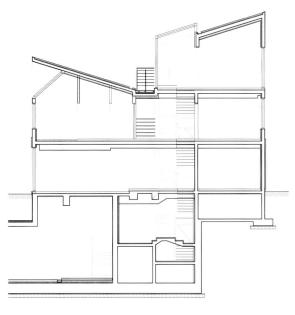

First floor plan

Section

■ Type-D

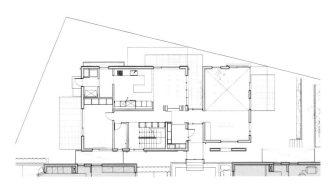

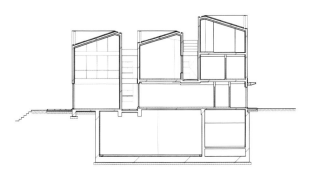

First floor plan

Section

Länsrådet住宅 **Länsrådet**

Location Stockholm, Sweden
Size 118㎡ x 14 houses
Architecture design Kjellander +
Sjöberg
Partners in charge Ola Kjellander,
Stefan Sjöberg
Team Maria Masgård, Karolina Sporre,
Erik de Vries, Ola Jonsson, Sylvia
Neiglick, Patrick Wüthrich
Interior photo Adam Mørk
Exterior photo Åke E:son Lindman

The objective in this project has been to establish an approach to the landscape of the Woodland Cemetery in Enskede south of Stockholm, a World Heritage site designed by Gunnar Asplund and Sigurd Lewerentz, and the built fabric of the single-family housing of Pungpinan.

This project lies across from the east entrance to the Woodland Cemetery. The vegetation surrounding the burial grounds, its towering tree trunks and mossy ground, with little undergrowth obstructing the views through the forest, continue on the other side of the road. To the back the site is closed off by rocky outcrops.

The fourteen houses form an arc around a shared outdoor space, sheltered from view and passing traffic. Seen from the open landscape of the cemetery, the scale of the project is subtly monumental, distinct. At the same time, the circular form blends in among the trees and does not dominate the view.

The landscape design, with components such as stone, bark, gravel, pine and birch trees, ferns etc. also works with the character of the site, a way to establish a connection with the Woodland Cemetery.

The details and materials used in the building have been chosen to articulate each individual unit rather than subordinating them to the whole scheme. This is clear from the street and the entrance. The second floor of each home has a large window that protrudes above the roofline. There's a well-

framed entrance court defined by natural-toned horizontal siding. The flat dark-brown colour of the rest of the building also helps visually blend the overall form with the surrounding forest, shifting our experience of it from direct and obvious to something everyday and intimate. The form first becomes clear on the back side, where it embraces, surrounds and defines a shared space, a community.

The lower level forms a kind of tunnel between front and rear—a space between two glass facades. As in the Sparsamheten development, it may be seen as a composite of a series of sub-spaces. It starts out at the street with the entrance court. After the kitchen and living area, the rear patio leads to a garden path. The balance between public and private shifts from one zone to the next: the character becomes progressively more private as you move from the street side toward the garden. The upper level is a completely private realm, with a small but useful multipurpose room with extra headroom and three bedrooms, a bathroom against the façade and a walk-through closet.

Site plan

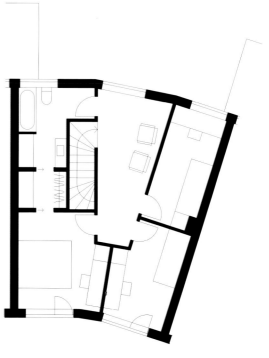

Second floor plan

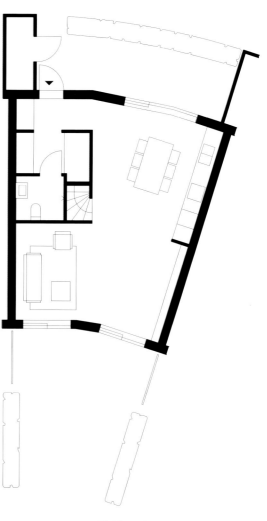

First floor plan

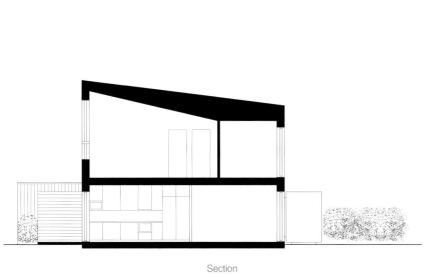

Section

Nord大街社会住房 Social Housing rue du Nord

Location Rue du Nord, Paris, France
Site area 350㎡
Total floor area 902㎡
Architecture design Charles-Henri
Tachon architecture & paysage
Photographer Kristen Pelou
Program 8 dwellings + 4 artist
studios + 1 retail area
Design participation Hugo
Clara(project manager), Jonas Houba,
Michael Aigner, Balint Szalontai

Constructed entirely in concrete, this building lies on reinforced foundations and is grounded on three fundamental objectives.
The first objective relates to the designation of Social housing. The word "social" is essential as it indicates the ability of a residential building to create social cohesion and exchange between its residents.
Sociability relies on the idea of recognition. At one hand recognition is, in a topographical point of view, a matter of discovering ones surroundings, including the perception of the surroundings from ones own apartment.
In this building even the smallest appartments have a variety of views to the outside. The diversity of the views provides a better understanding of the surrounding environment.

At the other hand, recognition does also include the recognition of others, to be a part of the city and to interact with the city. Every housing unit expresses itself in the façade and is quickly recognized in the neighbourhood as "the one with the large window", "the one with the terrace" or "the one on the roof".
Each resident becomes hereby an essentiel part of the building itself through his housing unit, and by this also a part of the neighbourhood and the city. This is already a form of social recognition.
The second objective is the uncompromising work with the plans of the housing units, which were the first priority during the entire work. There can not be any compromises.
A pleasant house is not only a spacious house, it's a house that

offers more. More space than it has surfaces, more functions and habitability.

In order to achieve spatial expansion and variety, diagonal views were created to provide transparency and complexity, sequences of continuously spaces, and a wide range of different interiors. Balconies and terraces were designed as directly extensions of the indoor livingrooms, and to be spacious enough to become comfortable and useable outdoor livingrooms.

A pleasant house is also, simply a house with natural light in every room, including kitchens and bathrooms.

As we found it improper to place the bathrooms on the facade side of the housing units, glass fanlights were used in order to provide indirect natural lighting.

Finally, a pleasant house, is a house that shows that the architect has been concerned about the details : the tiles for the bathrooms, the form of the sinks and to choose just the right faucets — and yet always with an awareness of the context and the financial latitude.

The shared spaces are spaces shared between neighbours who provide a common identity. The shared spaces were paid much attention, in the same way as each housing unit. A variety of ways to realise the different details and solutions were used, to show that even with a small budget, it is possible to pay attention to, choose and design each element. In other words, a great attention was payed in all scales of the project.

Moreover classical details has been reinterpretated as a part of the project. For example, the traditional Haussmanian staircase carpet has been here turned into a black painted lining on the raw concrete stairs and its white negative lines the underside of the flight of the stairs.

This brings us to the last but not least important objective: the act of architecture is a plastic act.

Our work focuses on the relationship between a space and the construction of this space.

The methods of construction, the materials, the forces, all these take part of the act of architecture.

This is why concrete was chosen, as it works in tension and compression at the same time.

Its implementation without visible joints gives the impression of a monolithic mass, and turns the building into a unit from which nothing can be extruded. The building is all about structure.

Its design reacts to the context of the street.

Setting back of certain parts at the corner of the façade contributes to make the building rise and appear more delicate, and it reminds us at the same time of the particular urban situation.

Further down the rue du Nord, the building adapts the surrounding volumes. While on the ground floor the artists studios front doors are aligned with the property line, the upper floors set back in order to create a longer distance to the building on the other side of the street, as the street is only 4,80 meters wide.

To accentuate the sculptural appearance of the building, every detail has been designed to reduce the distance between structure and space. For example, exterior railings were designed as simple glass sheets attached with lateral U-form steel profiles, the windows were conceived in order to include the shutters into the masonry, instead of the traditional steel windowsills, a concave drain has been shaped into the concrete

to collect rainwater and evacuate it thanks to a simple stainless steel drainage, and all the ironwork has been embedded directly into the concrete without any platinium fixations.

The traditional finish is no longer necessary as it has been conquered by the clear structure and materials

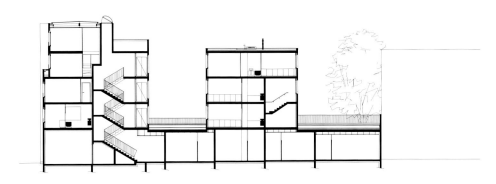

Longitudinal section

1. Living
2. Kitchen
3. Room
4. Bath

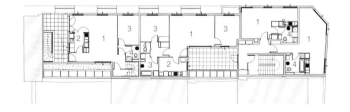
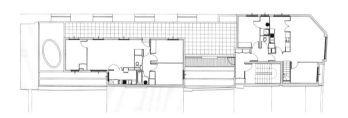

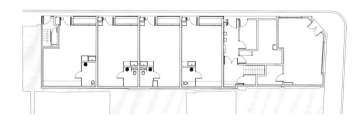
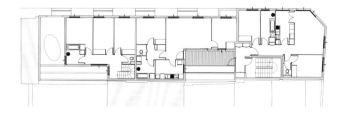

Floor plan I

Floor plan II

POMARANč公寓住宅 Apartment house POMARANč

Location Vinohrady, okres Bratislava III, Czech
Site area 351 m^2
Total floor area 597 m^2
Number of flats 3
Architecture design (AOCR) Ateliér obchodu a cestovného ruchu s.r.o.
Design ing. arch. Michal Vršanský
ing. arch. Vladimír Vršanský
Photographer Michal Vrsansky
Client CERSA s.r.o.

"POMARANč" house is situated on an interesting lot located between the main train station and the vineyards. The building itself occupies almost the whole site due to our aim to maximise the use under restricted conditions.

Sloped character of the area enables an access to the building by car or by pedestrians both from the street but from different levels.

In spite of practical use of the site the house benefits from various features bringing the standard of living in it to another level.

Three storeys high entrance hall (four storeys considering the rooflight), accessed from the street or from the garage contains central stairs with two fights and doors to each of the flats.

In the case of all living areas the window sills are brought down to the floor level to enable better connection of interior and exterior.

Duplex apartment situated on the upper storeys avails of open space with large glazed areas, rooflight over the staircase and secondary lit bathrooms via translucent walls.

Roof terrace on the top floor, interconnected with the living area and including jacuzzi, is taking advantage of the view of the city.

"The Orange" represents symbiosis of the "pragmatic house" with maximum use of space to clients benefit and "romantic house" in terms of individual enunciation.

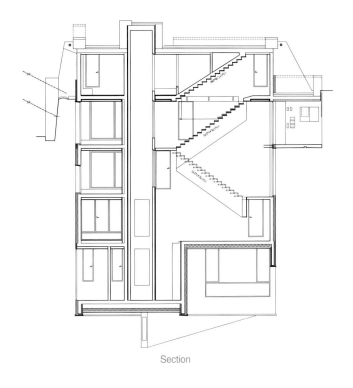

Section

1. Living
2. Kitchen
3. Terrace
4. WC
5. Room 01
6. Bath
7. Room 02

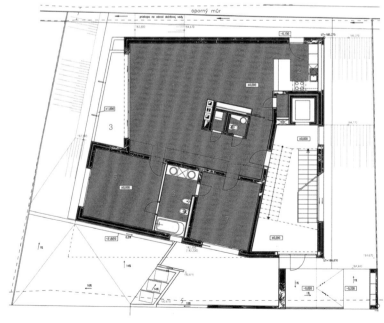

Ground floor plan

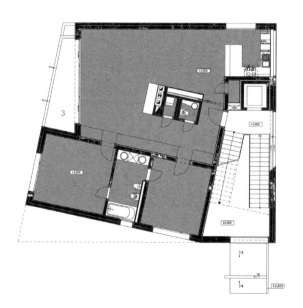

First floor plan

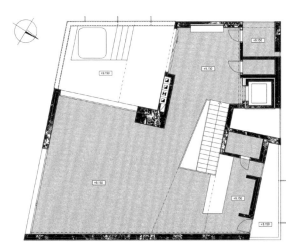

Thrid floor plan

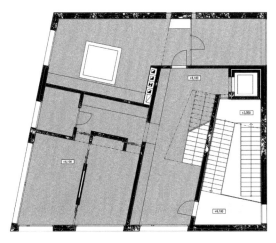

Second floor plan

Jagokdong Motgol市镇住宅
Jagokdong Motgol Town

Location Seoul, Korea
Site area 330㎡(Typical flat),
326㎡(Typical courtyard),
315㎡(Typical triplex)
Bldg. area 163㎡(Typical flat),
162㎡(Typical courtyard),
157㎡(Typical triplex)
Total floor area 519㎡(Typical flat),
545㎡(Typical courtyard),
525㎡(Typical triplex)
Architects KYWC Architects / Kim
Seung Hoy + Kang Won Phil
Design participation Sohn Seok-hoon,
Jin Seong-il, Kang Dong-ki,
Kim Jung-yun

This project is in Jagok-dong, Seoul Metropolis. The project is suggesting new types of urban residence with higher density by creating a single independent residence for two households. The design suggested three types, which are; courtyard type, flat type and triplex type.

The courtyard type has its external space and internal space integrated. It gives extensive space-feeling and enables the locating of various spaces. The traditional space-structure that links yard, room, road and gate was adopted. The house has cross-section that meets with topography and the city context by active utilization of sloped land. Triplex type and flat type (stacked flat units) were suggested together with the courtyard type.

Stone was consistently used in all three types. As auxiliary materials, wood and metal were used to metaphor the rich space-character and let the house have various scales and textures. The house is for two households; however, the design was to let the house be recognized as a single house. It was possible by making the facade asymmetric. The metal roof and eaves in roof-tile color work as devices responding to climate and external space, like in traditional Korean architecture. They are also the architectural language welcoming the family members coming home.

1. Living room
2. Dining & Kitchen
3. Master bedroom
4. Bedroom
5. Multi-purpose room

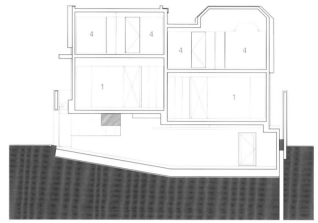

Section E-W (Typical courtyard)

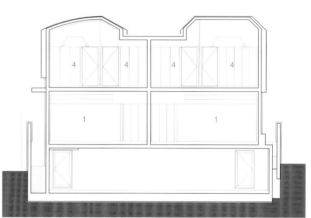

Section E-W (Typical flat)

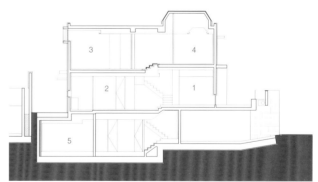

Section S-N (Typical flat)

■ Diagram

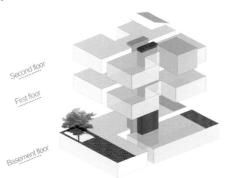

Second floor
First floor
Basement floor

Typical courtyard

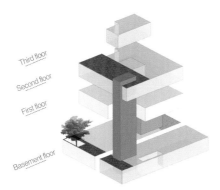

Third floor
Second floor
First floor
Basement floor

Typical triplex

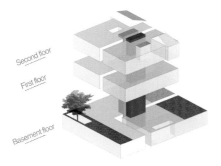

Second floor
First floor
Basement floor

Typical flat

1. Entry
2. Living room
3. Dining & Kitchen
4. Courtyard

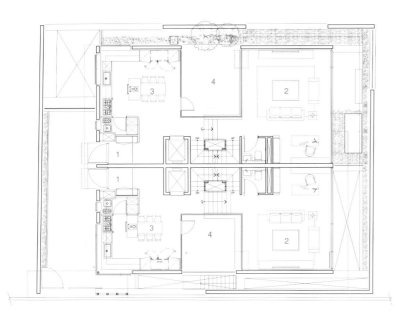

First floor plan (Typical courtyard)

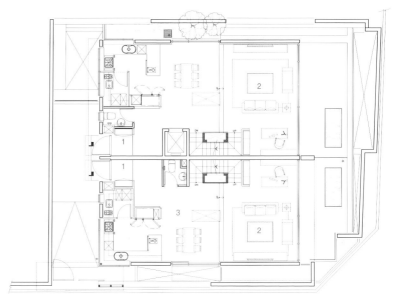

First floor plan (Typical flat)

达拉斯碗形住宅 Dallas Bowl Housing

Location Dallas, USA
Bldg. area 47,000㎡(housing, retail)
Architecture design ON OFFICE
Design participation Joao Vieira Costa,
Leon Rost, Ricardo Guedes,
Joao Amaro, Nitsan Yomtov_
USA(collaborator)

The overall strategy of Dallas Bowl is to incorporate the urban, environmental, programmatic, and social aspects of this residential block into a holistic, fully-functioning organism, where all elements of the built and natural environment synergize in a symbiotic relationship.

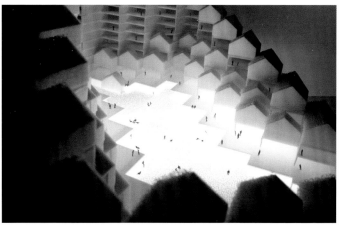

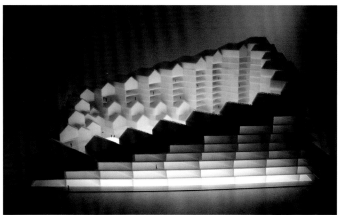

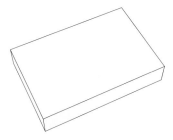

– We begin with a "full cake" of the required program.

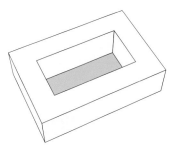

– We excavate a courtyard to let in light, to create a destination, and to foster a sense of community.

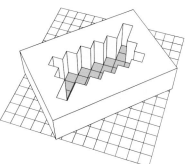

– The block is reoriented towards the cardinal directions.

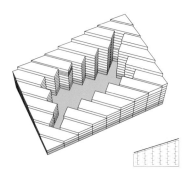

– The block is tilted towards the south sun, to harness the maximum amount of solar energy. The block is sliced into detached rows.

– A blanket of pitched/roof houses are added to the top layer. The south face is clad with PV, generating power for the entire building. The north face harbors plants and collect rain.

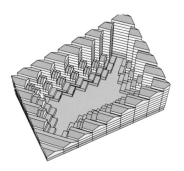

– The courtyard is articulated to bring the scale down to that of a human, and to add potential to commerce, farming, and public life.

Form generation

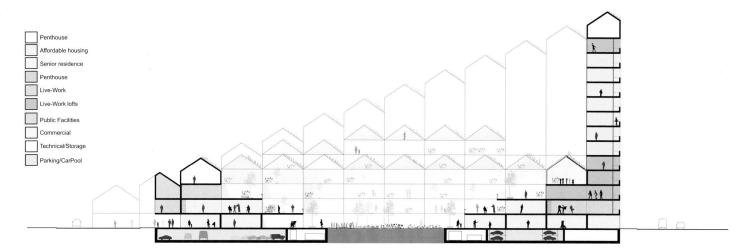

Penthouse
Affordable housing
Senior residence
Penthouse
Live-Work
Live-Work lofts
Public Facilities
Commercial
Technical/Storage
Parking/CarPool

Section

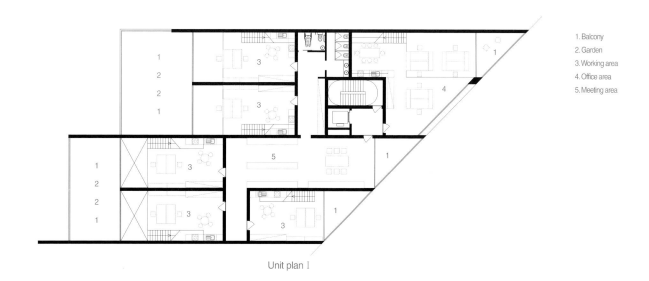

1. Balcony
2. Garden
3. Working area
4. Office area
5. Meeting area

Unit plan I

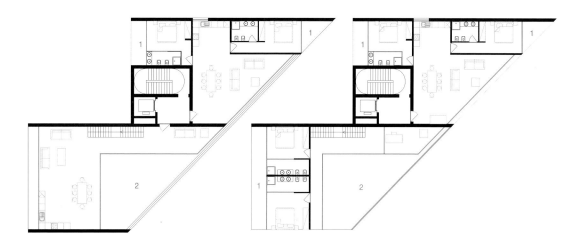

Unit plan II

KMM3公寓 KMM3 Apartment

Location Tokyo, Japan
Unit 9 units
Architecture design ISSHO Architects
Collaboration Shuji Tada(Structure)

This is a 9 units apartment building in Tokyo. By introducing an intermediary space at a window between inside and outside, its design redefines an experience of living in a typical low-rise housing in a highly dense area of Tokyo, where scenery from room and distance among buildings is often not pleasant. The intermediary space has variation in depth, which could diffuse a sense of distance between interior and outdoor, thereby an occupant comfort is enriched.

米歇利特50号住宅 Michelet 50

Location Anzures- Mexico City, Mexico
Site area 405.98㎡
Built area 1568.56㎡
Photography © Sandra Pereznieto,
María Luz Bravo
Architects Dellekamp Arquitectos +
Gerardo Asali
Project Leader Ignacio Méndez
Team Aisha Ballesteros, Jachen
Schleich, Pedro Sánchez

The apartment building is situated in the centre of Mexico City, in the district of Anzures, indicating the urban character of the plot. According to the requirements of the client, we designed a building which features different apartment typologies divided into three blocks, giving more square meters of facade to each side. Common areas of the apartment community are contained in a single space looking outside to three facades. The building stands away behind the street line, giving space to create green zones in an urban district. Each of them surrounds and enters into the volume to become service areas. This idea became the main driving force behind the interior space design. A glass facade covers the irregular wall of the

building to provide a unique appearance, also giving a character of unlimited connection between interior and exterior space. Just the horizontal framing marks the different levels of the construction.

Meanwhile we were developing the project; the Mexican artist Jeronimo Hagerman installed one of his plastic art works in the Museum Sala de Siqueiros of Mexico City, using veining plants perfectly the ones we were thinking for the green areas. In collaboration with the artist, we kindly could insert the plants in the project, once finished the exposition; giving them a new utility.

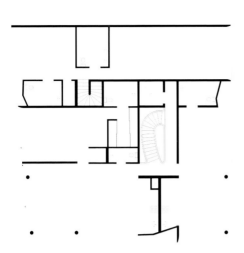

Floor plan

南安普顿米尔布鲁克Porlock路住宅
Porlock Road Millbrook Southampton

Location Southhampton UK
Unit 11 residences
Tenure mix 9 social rent, 2 shared
ownership
Site area 0.2ha
Architecture design architecture plb
Developer Spectrum Housing Group
Contractor ISG Pearce
Photographer Timothy Soar

The Site and Design Solution

The site sits at the edge of an established 1960s suburban context, with a large area of open space to the north and west. The site is flat and is dominated by a large number of protected trees. Existing dwellings within the area consist mostly of 2 storey houses and 3 storey blocks of flats.

The proposals were formed by a combination of the kit elements in a particular manner to suit this site. We have taken care to ensure that the design functions and appears as a site-specific solution. This has been achieved through a careful arrangement of elements, relating to specific contextual issues such as neighbours, trees and views, and through the treatment of the building form, public realm and material selection.

The mix of building scales within the area was taken as a cue for the proposals. Roofs are pitched to reflect the surroundings.

2 storey elements are placed adjacent to existing houses to reduce the impact and existing trees retained to provide screening. The proposals sought to achieve a bold architectural form, creating a strong sense of identity and place, whilst remaining sympathetic to the surroundings. A contemporary language has been used, "softened" by careful articulation and mix of materials. Each unit is clearly differentiated, with distinctive entrances enhanced by generous canopies.

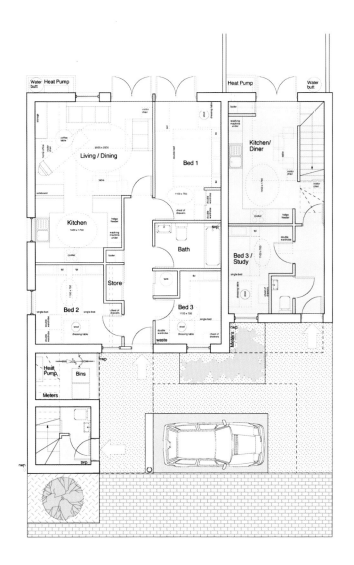
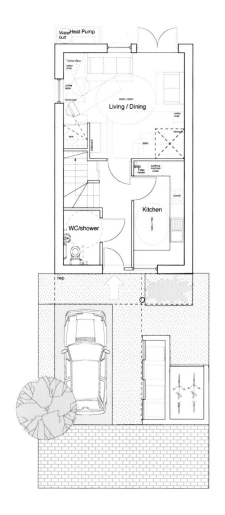

Floor plan

Layout plan

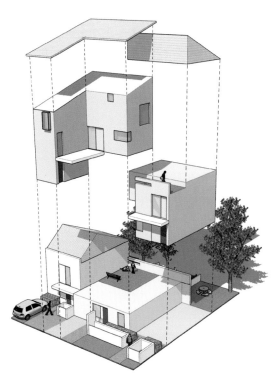

Exploded diagram

Front elevation

Rear elevation

黑色住宅 The Black House

Location Utrecht, Netherlands
Function Atelier+ 6 Apartements
Project area 1,100㎡
Architecture design Bakers Architecten
Photography Maarten Noordijk (foto@ maartennoordijk.nl)
Frank Stahl (f_stahl@gmx.de)

In Utrecht's museum quarter, just south of the city centre, there was for many years a vacant plot on the corner of Lange Nieuwstraat and Vrouwjuttenstraat. This site in the midst of historical buildings is now occupied by "Het Zwarte Huis" (The Black House), a complex containing six apartments with semi-underground parking and the new premises of Bakers Architecten.

The streetscape is characterized by heterogeneous, lot-by-lot development with distinctive corner buildings. Het Zwarte Huis is a contemporary addition to the existing urban fabric, in which the notion of "living above work" has been accentuated by placing the dwellings in a solid volume on top of a glazed podium.

Lange Nieuwstraat begins at Domplein and runs via a gentle curve to the Centraal Museum. The site lies at the mid-point of the curve from where there is an overview of the entire street.

This unique vantage point is fully exploited with a large bay window.

An internal courtyard has been created by placing the black volume parallel to the Lange Nieuwstraat. This volume also contains the various means of access for the complex as a whole. The semi-underground car park is reached via a car parking lift, while a communal staircase leads to the walkways along which the apartments are situated. The wide walkways also serve as outdoor space for the dwellings.

Het Zwarte Huis was constructed using 55-centimetre-long "Kolumba" bricks. The apartments facing Vrouwjuttenstraat have a white rendered facade. The party walls on this side form a cantilever on Vrouwenjuttenstraat, thereby relieving the podium facade of any structural function and allowing it to be entirely of glass.

© Frank Stahl

© Maarten Noordijk

© Frank Stahl

© Maarten Noordijk

© Frank Stahl

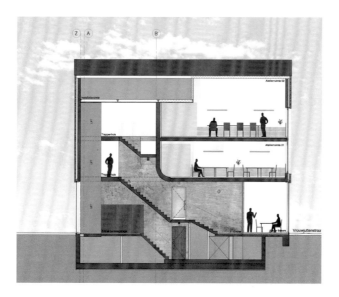

Section

© Maarten Noordijk

© Frank Stahl

Yangji Balt住宅 **Yangji Balt House**

Location Gyeonggi-do, Korea
Site area 660㎡
Bldg. area 132㎡
Total floor area 321㎡
Architects KYWC Architects /
Kim Seung Hoy+ Kang Won Phil

The Balt house is the largest residence in this whole housing complex. The three houses built in the sites of different forms and conditions were designed based on the same motive yet with unique characters. The exteriors of the entire complex are finished with bricks and share a common design language. The layout of the exterior spaces, the composition of eaves and walls, and the pavilion on the third floor reflect the beauty of a Korean traditional house, while the interior volume and the continuous spatial flow that starts from the living room to the dining room and kitchen show the contemporary lifestyle. The spatial system inside, which is divided according to use, reveals itself through the appropriately segmented exterior. The separated volumes made of bricks are of a comfortable scale,

and the space between the interior and exterior is enriched by several layers of eaves, walls and a balcony. The porch canopy, the eave attached to a window, and a low wall show how this house responds to the climate and communicates with the outer world. The posture of the residence, which has been designed to have a close relationship with its surroundings, gives a sense of uniqueness to the three houses located in the sites of different conditions.

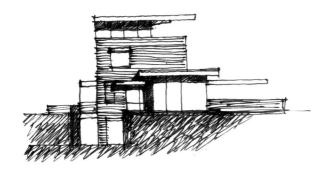

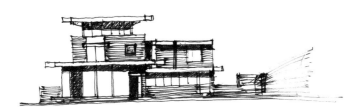

1. Entry
2. Living room
3. Dining & kitchen
4. Hallway
5. Bedroom

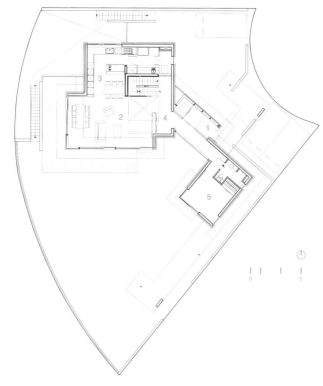

Typical floor plan I

Typical floor plan II

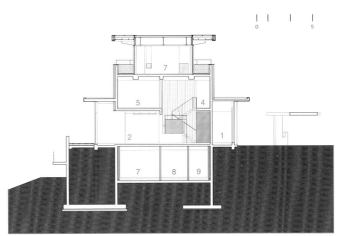

Typical section I

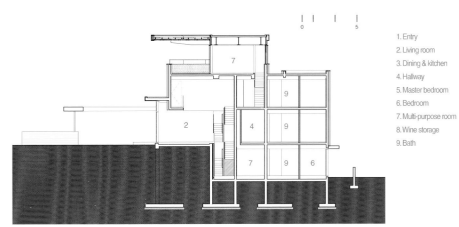

Typical section II

1. Entry
2. Living room
3. Dining & kitchen
4. Hallway
5. Master bedroom
6. Bedroom
7. Multi-purpose room
8. Wine storage
9. Bath

DE住宅 **House DE**

Location Clearwater Bay, Hong Kong, China
Size 4,300 sf
Design davidclovers
Design participation David Erdman, Clover Lee, Mui Fuk Man, Jason Dembski, Yvette Herrera, Rathi Subramanian, Lam Pui Wing, Spencer Mak
Photographer davidclovers, Almond Chu
Type Two Units Townhouse

House DE is an "infill" townhouse, spectacularly sited on a hillside above Clearwater Bay, Hong Kong. Combining two existing units into one, the design uses the volumes of three staircases to blend, burrow and interlock spaces vertically across four floors. Each "interaction" is materially monolithic, using stone, wood and a series of delicate aluminum fins. Defined by these fins, the texture and form of the lantern-volume subtly changes shape and depth, casting shadow and emitting light in different ways throughout the day. Each stair-volume pries open the house vertically and horizontally, pulling in daylight and emitting artificial light. Thickening the existing building enclosure and stretching it across the front and rear, the bedrooms and new master suite on the upper floors are protected from the elements, yet open up to views of the natural surroundings. Towards the South, the facade thickens and torques, providing shade for bedrooms and balconies; while on the North, the facade transforms into a garden trellis for an outdoor dining terrace.

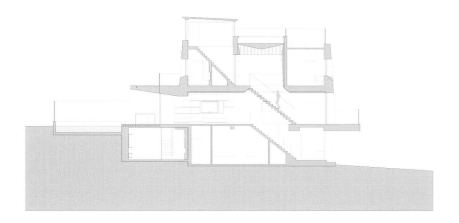

East section

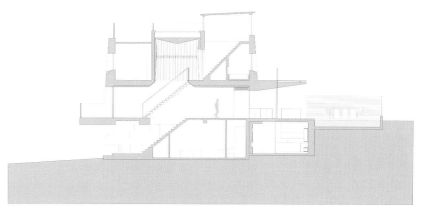

West section

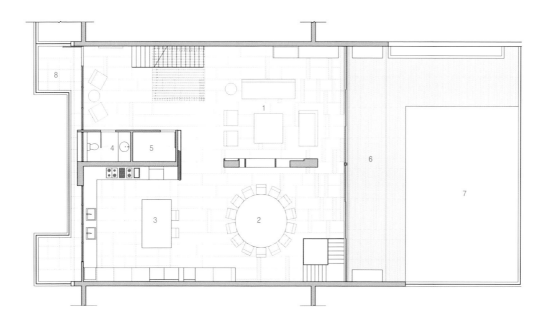

1. Living Room
2. Dining Room
3. Kitchen
4. Powder Room
5. Sundry Room
6. Outdoor Dining
7. Garden

First floor plan

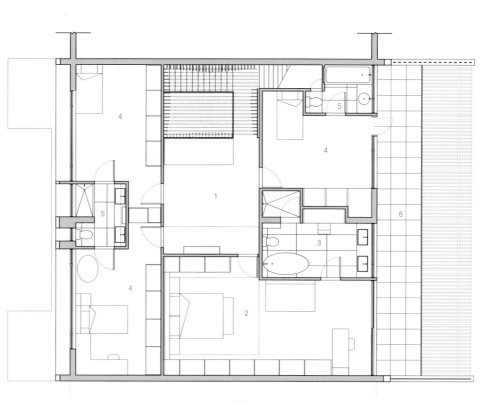

1. Family Room
2. Master Bedroom
3. Master Bathroom
4. Bedroom
5. Bathroom
6. Balcony

Second floor plan

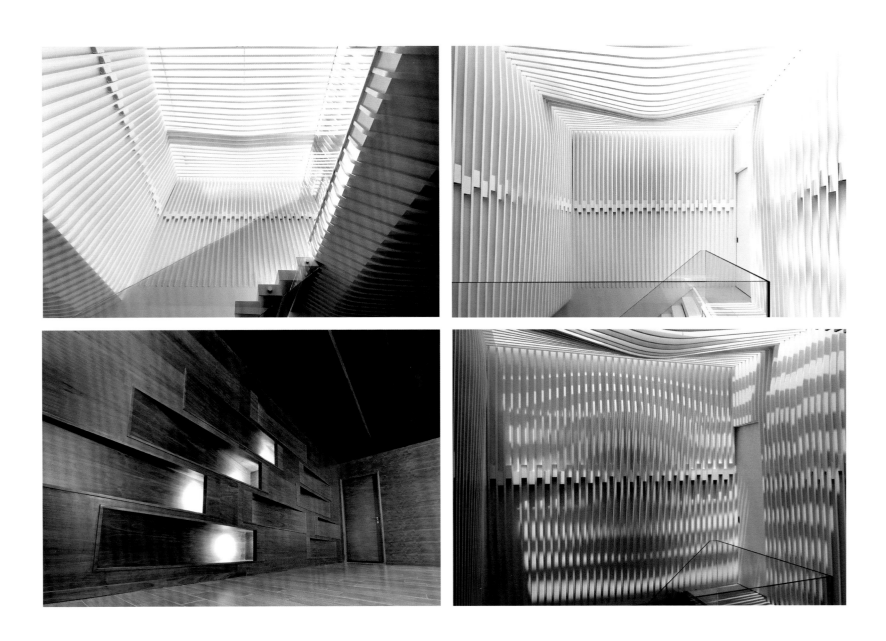

Kaminoge住宅 **Kaminoge House**

Location Setagaya-ku,Tokyo
Site area 562.65㎡
Built area 288.33㎡
Architecture design NAOYA
KAWABE ARCHITECT AND
ASSOCIATES
Designer Kawabe Naoya,
Tsuji Masashi ,Taoka Hiroyuki

A moderate distance is kept outside by assuming the town house form. I thought that an independent approach to each unit is connected at the outside environment with planting and the chance to touch. In addition, the town house was not made overbearing an one volume. We were made in the south-north through by using "Division" and "Gap". It is tried to take the outside environment indoors by apply the daylight and ventilation from the window opened to the beast position and made each unit section three-dimensionally. Then, the throb feeling in a popular apartment without was able to be given.
At the same time, it aims at the shape that merges without shutting though the interplay of the glance of the unit.

South elevation

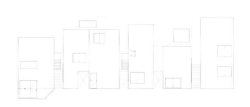

North elevation

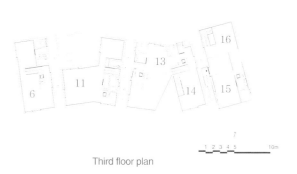

Third floor plan

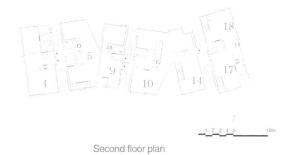

Second floor plan

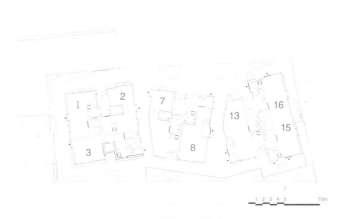

First floor plan

Section

Krautgarten的住宅开发
Housing Development in Krautgarten

Location Vienna, Austria
Apartment Units 17 units
Underground Parking
Spaces 17 cars
Gardens / Terraces 500㎡
Gross floor area 2,705㎡
NUA 1,590㎡
Architecture design Caramel
Architekten
Photographer Hertha Hurnaus
www.hurnaus.com

The property lies in a green zone at the periphery of vienna and is surrounded by small houses and small apartment complexes. The atmosphere that future tenants associate with this site is one of "living in green surroundings", ideally: one's own house and garden. The intention is to also convey this atmosphere in the housing project itself. The individual apartments are designed as separate houses with multiple floors and various views of the surroundings. The individual units are interlocked and fit together to create the overall structure. Although the krautgarten project is a four-storey building, each individual living unit was designed as self-contained as possible. Each apartment, therefore, has its own floor plan and the various loggias and balconies allow the outside observer to recognize each unit as a distinct "house". Private yards, roof gardens, loggias, and balconies provide each apartment with various levels of direct contact to the green surroundings – like a house surrounded by grass and trees.

The individual internal structure of the building gives the outer shell a playful character as a positive atmosphere in a functional unit "for living". This is emphasized by the haptic impression of the very course plaster of the building exterior and the cheerful choice of colors.

Urban development

From an urban development perspective the site is located in a very heterogeneous environment. The corner of the building facing the public traffic area assimilates the main directions of the adjacent buildings. A green zone is produced along the street as a buffer between public and private space.

Architecture / organization

The main target group for this housing development with "green" surroundings at the periphery of the city comprises young families and the like. More than half of the apartments have their own gardens; the rest have terraces or loggias.
In addition, the apartments – mainly terrace-house-style duplex

units – feature a high degree of individuality. The differences in the apartments can also be seen in the exterior design and the façade. The loggias ("eyes") indicate the living rooms; the smaller windows arranged in varying patterns mark the bedrooms. Located virtually at the "heart" of the building, at the geometrical center of the ground floor, and adjacent to the stairwell is a multipurpose zone and children's play area with a sandbox and monkey bars.

Common zone (play and meet)

The common zone consists of a wooden surface including a large sunken area for playing with sand and a vulcanized rubber play area with a hill. In addition, there is a long multipurpose element good for both sitting and lying on. Despite its small size, the area is intended as a space where residents can meet, talk, and play together.

South elevation

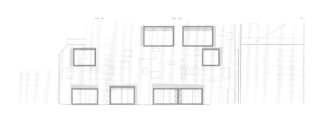

North elevation

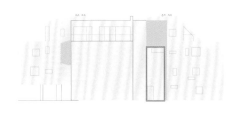

East elevation

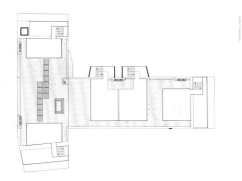

Floor plan I

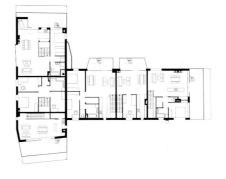

Floor plan II

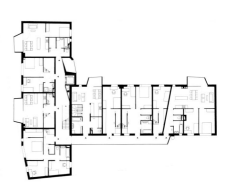

Floor plan III

梅耶尔住宅 House Meijer

Location City of Almelo,
Netherlands
Project area 370㎡
Architecture design Van der Jeugd
Architecten
Design participation Paul van der
Jeugd,
Ruud van der Koelen, Mirjam
Wiggers
Photography Ruud van der Koelen
(www.vanderjeugd.nl)

The house is located in the newly developed residential area "Hegeman" in Almelo. Within this plan, 9 lots were available for free sale. These plots are well positioned within in the district: adjacent to the canal, with a clear view of woods and meadows. Houses on these lots should to be unique and innovative, with great architectural quality.

The plot acquired by the client is located along a green plot which is kept open, creating a vista from the residential area behind the plots to the surrounding landscape of hedgerows, fields and canals.

During the first visit by the client to the architect, they were very interested in the in situ-cast concrete work office of the architect. This led us to design a concrete dwelling: spatially strong, with an aesthetic minimalism put through to the last detail. Strategically rainwater drains, piping and electrical wiring were eliminated from sight, resulting in an almost Spartan and graphical appearance.

Knowledge and experience of in situ-cast concrete were used in both design and building process. This has led to a series of thoughtful and ingenious details. Also the building process was well controlled and directed by Van der Jeugd Architecten.

The dwelling was tailored to the needs of the client and the specificity of the location. When organizing the agreed housing program, the particular views and the specific light fall were of

importance. This has led to a three-storey dwelling house and four specific spatial levels.

The design of the ground floor (split level) is such that a series of area's were created with its own character, spatially and visually in contact with other area's: intimacies with a minimum of visual and spatial separation.

Much care was taken to the staircases and outdoor spaces. Daylight openings are strategically positioned in the house, optimizing the experience of light and space. The balconies and outdoor spaces are positioned in such way that all the interesting views can be experienced, and are important in the appearance of the house.

Meyer House is more than one dwelling. It is a modest but extremely life-proof living machine, which actively enriches the everyday experiences of its residents, increases the quality of the site and particularizes a sense of space.

1. Kitchen
2. Dining
3. WC
4. Living Room
5. Terrace
6. Entry
7. TV Room
8. Study Room
9. Carport

Basement plan

Ground floor plan

Section B-B

1. Bedroom
2. WC
3. Balcony
4. Bathroom
5. Corridor
6. Bedroom

First floor pan

Second floor plan

Roof plan

Section A-A

悬崖住宅 The Cliffs

Location Brisbane, Australia
Bldg. area 3,550㎡
Total floor area 380㎡
Architecture design Ellivo Architects
/ Mason Cowle, Scott Whiteoak
Interior design Tanya Zealey
Photographer Scott Burrows

The Cliffs

Brisbane has flourished in recent years with a flux of development that takes advantage of the Brisbane River and the amenity that comes with living close to the city heart. This has created a demand for smaller scale, more boutique developments as an adjunct the many large scale towers. "The Cliffs" is an example of the evolution of this new apartment typology. The design objective is to provide a high quality living environment which combines the convenience of inner city living with the quality and attention to detail of an individual house. The result is seven residences over four levels, with private lift access and city views to the north. Living spaces flow to the generous decks through bifold doors. The two apartments per floor arrangement allows for full cross ventilation of the apartments, reducing the need for air conditioning. Common facilities for the residents include a large pool deck area, gym, bbq and associated deck spaces. The penthouse includes its own roof top pool and bar area. Acoustic amenity has been carefully considered to lessen the noise effects due to the proximity of the Captain Cook Bridge. Consideration has given to acoustically protecting outdoor spaces as well as indoors spaces. Apartment terraces have been partially enclosed with curtain walled screens, noise deflecting handrails and sound absorbing ceilings. Externally, stone cladding combined with Alucobond panels and Evergreen glazing articulate the facade. The material selections are low maintenance and understated.

Penthouse

The Cliffs Penthouse offers living and outdoor spaces that are a haven for relaxation or a sensational entertaining venue. Spread across two levels, atop the Kangaroo Point cliffs, the sweeping views to the city and storey bridge were captured by the arrangement of furnishings designed to enjoy the Cliffs lifestyle.

North elevation

Street elevation

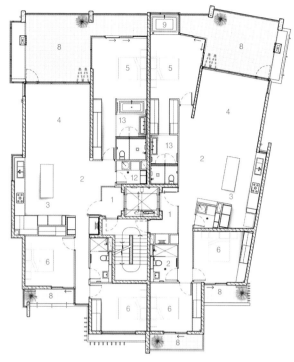

Typical level

1. Foyer
2. Dining
3. Kitchen
4. Lounge
5. Master bed
6. Bedroom
7. Bathroom
8. Balcony
9. Spa
10. Deck
11. Media room
12. Laundry room
13. En suite
14. Pond

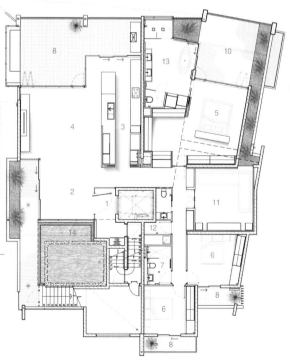

Penthouse floor plan

Dongtan Chungdo Solium联排住宅
Dongtan Chungdo Solium Town-House

Location Gyeonggi-do, Korea
Site area 18,214㎡
Bldg. area 6,327㎡
Total floor area 17,224㎡
Interior design D.I.A.N Architecture Co. /
Suh Yeun-ju
Photographer Kim Jae-youn

Urban landscape and the life of residents within the residential complex, is largely dependent on the owner's development direction. The beginning point of this project, based on the pride of mutual coexistence, and realization of new residential culture — instead of economy or profit — is therefore man and nature, not the architectural form or economy. Through 70~80's rapid industralization, our urban landscape with diverse alleys and roads, become filled with uniform apartment and asphalt roads. The residential culture of us which was rapidly changed and grown, get spoiled by vanity and ostentatious display. Therefore, our house should return to its original position, i.e. to the daily space of people, instead of means of profit or wealth. Narrow and close sense of space with full floor space index,

and its uniformization by conveniency and economy should be sublated. Instead, the spirit of this complex is to provide a space where residents's personality and time could be reflected, through the room and freedom of emptiness. To create a residential space resembling the peace of tree, and the place where man do not conflict with nature, or destroy its forest by the reason of urbanization. Sound man in a sound building, our children growing in the peaceful village of nature, would resemble its nature, and be strong and generous. This is our duty in this land.

• A-type

• B-type

• A-type

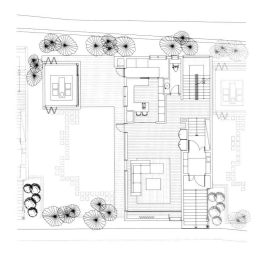

First floor plan

Elevation

• B-type

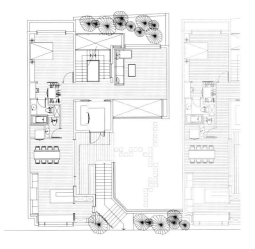

First floor plan

Elevation

图书在版编目(CIP)数据

155个居住设计：英文/韩国建筑世界出版社编
. 一大连：大连理工大学出版社，2012.2
　ISBN 978-7-5611-6629-1

　Ⅰ.①1… Ⅱ.①韩… Ⅲ.①住宅－建筑设计－世界
－图集 Ⅳ.①TU241-64

　中国版本图书馆CIP数据核字（2011）第238681号

出版发行：大连理工大学出版社
　　　　　（地址：大连市软件园路 80 号　　邮编：116023）
印　　　刷：精一印刷（深圳）有限公司
幅面尺寸：260mm×300mm
印　　　张：58
出版时间：2012 年 2 月第 1 版
印刷时间：2012 年 2 月第 1 次印刷
出　版　人：金英伟
统　　筹：房　磊
责任编辑：张昕焱
封面设计：王志峰
责任校对：王单单

书　　　号：ISBN 978-7-5611-6629-1
定　　　价：748.00 元（共 2 册）

发　行：0411-84708842
传　真：0411-84701466
E-mail: a_detail@dutp.cn
URL: http://www.dutp.cn